THE STORY OF LAVENDER

Sally Festing

To Nicky

Lavender fields at Caley Mill, north Norfolk

The Story of LAVENDER

Sally Festing

London Borough of Sutton Libraries and Arts Services

First published 1982

London Borough of Sutton Libraries and Arts Services
Central Library, St. Nicholas Way, Sutton, Surrey

Tel: 01-661 5050

ISBN 0 907335 05 5

DESIGN: SHIRLEY EDWARDS

Second impression 1984

Printed by John Bentley (Printers) Ltd., Todmorden.
A member of the Dunn & Wilson Group

ACKNOWLEDGEMENTS

My grateful thanks are due to Douglas Cluett and June Broughton of Sutton Libraries for help and extensive editorial assistance; to Mr. E. M. Montague, Mr. C. D. Brickell, Mrs. A. Head, Miss V. Lewis, Dr. R. A. Scott and Prof. W. T. Stearn for reading parts of the manuscript; and to Michael Harkin of Mitcham Library, Dr. Brent Elliott of the Lindley Library, Royal Horticultural Society, Miss Jones of the Pharmaceutical Society Library, and the staffs of the Library of the Royal Botanic Gardens, Kew; the Wellcome Institute Library; and Hertfordshire County Library, Local Studies Department. Thanks should also go to Mrs. Mavis Batey of the Garden History Society, Mr. M. J. Milchard of Tropical Products Institute, Mr. John Harvey and Mrs. Kay Seneki for help during the book's early stages, and to Valary Murphy for her sterling work in typing the manuscript.

Photographs were kindly lent by Miss V. Lewis of Hitchin, the Guildhall Library, Merton Library Services, and Yardleys of London, and are reproduced with their permission. The illustrations on page 41 are reproduced by kind permission of the Oxford University Press and the Sun Publishing Company, and appear in the *Oxford Dictionary of Nursery Rhymes* by Iona and Peter Opie; those on pages 67 and 68 are taken from G.W.S. Piesse's *Art of Perfumery,* and those on pages 13 and 14 are from *The Natural History of Lavender* by Baron Frederic de Gingins-La Sarraz.

The illustrations on pages 22 and 29 are reproduced by kind permission of the Lindley Library, Royal Horticultural Society; the first (p.22) is from Adam Lonitzer: *Kreuterbuch,* 1557 *via* Frank Crisp *Medieval Gardens,* London, 1924; the second (p.29) is from J. Brunschweik: *Das Distilierbusch,* 1521, *via* Anthony Huxley, *An Illustrated History of Gardening,* Paddington Press and Royal Horticultural Society, 1978. Other illustrations are from Sutton Libraries' Local Collection or were taken by the author.

Contents

Foreword

When we first moved to South London I became interested in lavender. This is commuter land, mopped up into the metropolis and planted with ubiquitous suburban cherry trees. People come and go; and one town, one village, slides almost imperceptibly into the next. But it was not always like this. Croydon, Beddington, Wallington, Cheam, Carshalton and Sutton used to be highly individual. One thing, however, they had in common: all grew lavender when, in the wake of modernisation, the innocent plant fled from Mitcham.

Through realisation of the past a sense of involvement is captured; you need to dig back a bit in order to belong. So I have talked with some of our older inhabitants and they can remember going out with a gun to catch a bunny for lunch, passing through waving grey lavender fields on the way. Indeed, the house we inhabited in Carshalton's Banstead Road once stood on lavender fields, and neighbours had clauses written into their deeds prohibiting them from planting lavender for commercial purposes, presumably to avoid competition with the growers.

Today our windows reveal a network of chimneys and tiled roofs on every side. But I am conscious of, and to some extent regret, the loss of the idyllic picture of this area, the site of the farm of Mr. Woods, conjured up in an old newspaper. "In every direction the low hillsides of the farm are swept with bloomy pastel tint of reapers in the fields. As the day wears on, the fragrance rises like incense in the air, wandering tribes of paper-white butterflies drift over the fields and in the clear depths of blue sky, larks discant the joy of life."[1]

Everyone knows lavender. I have no need to describe the curious little downy plant that sends the bees into such a summer tizzy.

Throughout England, its pin-cushion tufts throb with scented heads: it is familiar garden fare. Nor need I describe the sharp, clean fragrance that perfumes our hair, our hands, our clothes and faces. Arab women once used its oil to add lustre to their swinging tresses, and Roman ladies preferred its perfume to the precious nard of India. During the chaotic days of the Dark Ages, when the Church steadily grew, and offered hope, it was monks who preserved knowledge of herbal lore and fostered the gentle art of gardening. They planted and primped lavender behind monastery walls. In the high Middle Ages, the chivalrous years, it was grown in rectangular beds inside semi-fortified moated granges in Britain and on the Continent.

By the 16th century, lavender water was being distilled in many English country houses; and in the 17th it was used to disguise London's stinking streets and to ward off the plague. Contemporary medical practitioners carried, at the end of their walking sticks, a cassolette filled with aromatics, which they held to the nose as inconspicuously as possible when they visited the sick. Lavender was even used for making a spirituous tincture, enjoyed by the raffish, according to Sir James Smith (1759-1828) as a source of alcoholic beverage: "a popular cordial, very commodius for those who wish to indulge in a dram, under the appearance of elegant medicine".

In France the art of perfumery was raised to its highest level. Lavender is still gathered by Alpine farmers from arid hills and uncultivated plains, where its silvery leaves mingle with the sombre green of the heath. You might expect the plant to secrete its aromatic oil more freely in a hot, dry climate, but, in fact, lavender, rosemary and peppermint cultivated in Britain produce oil every bit as good, as fragrant and abundant, as that from anywhere else in the world. Three hundred acres of lavender were grown in and around Mitcham in the mid-nineteenth century, and the oil produced there realised six times the price of its French counterpart.

Perhaps these were its balmy days in this country. Since then it seems to have dwindled to about 100 acres in North Norfolk. But are there any lavender fields left in South London? Is it grown anywhere besides Norfolk on a commercial scale? Suddenly I wanted to know more about the plant. Like a faded sampler it conjures up memories of the past – of sun, warmth, bees; and more leisurely days.

1
Origin of the Name

Most people think lavender was named from the Latin, *lavando,* part of the verb *lavare* – to wash, because the Romans used it to perfume their baths. But an old herbalist called William Turner had a slightly different explanation: "because wyse men founde by experience that it was good to washe mennis heades with, which had anye deceses there in";[2] and a certain Victorian doctor argued that its earliest spelt form was livendula, which was more likely to be connected with the Latin *livere* – to be livid or bluish.

At any rate, lavender has continued to symbolise cleanliness and purity. There is a reference in the *Liber Niger,* or Black Book, of Edward VI to a "lavender man" authorised to obtain from the Spicery "sufficient whyte soap tenderly to wasshe the stuffe from the King's propyr person";[3] and he was the launderer, not a perfumier nor a lavender seller.

William Langham advised: "Boil it in water, wet thy shirt in it, dry it again and wear it . . .";[4] while the gentle Izaak Walton, lover of the countryside, wrote during one of his angling expeditions almost a century later: "Let's go to that house, for the linen looks white and smells of lavender, and I so long to be in a pair of sheets that smell so".[5]

On another occasion, he took his friend Venator to an "honest ale-house" where the room was clean and lavender stood on the sill. It was customary at the time to place pots of lavender, rosemary, and other scented foliage plants at the windows of village inns.

Another fifty years later, Keat's Madeline dreamed tranquilly between sheets "smooth and lavender'd"; whilst her lover, Porphyro, prepared to whisk her into a stormy St. Agnes' Eve.[6] Leigh Hunt brings the idea of fragrance to matrimony: "pure lavender, to lay in bridal gown";[3] meanwhile, the old association of lavender with laundry continues. Even today, Italian housewives drape their washing over rosemary and lavender bushes to absorb the fragrance as it dries in the sun.

English lavender, dwarf, at Caley Mill, N. Norfolk.

2
Different Kinds of Lavender

Inconsistency over naming the lavenders has led to the same word being used simultaneously for quite different plants. To add to the confusion, they readily cross or hybridize, giving rise to a flow of new types; and botantists vary in their willingness to accept these newcomers as distinct species. This means that species claimed by one school of botany may be discarded, later, by a more purist faction. So the group appears to expand and contract on the whim of its classifier.

It was not until 1937 that a Miss D. A. Chaytor from Kew took the genus properly in hand. Using a classification based on a combination of two previous ones, she divided the lavenders into five sections, distinguished chiefly by differences of habit and anatomy. But these sections are by no means always clear cut, and intermediates which cannot be accounted for simply by hybridization seem to occur.

All the common garden and commercial plants belong to two sections, stoechas and spica. The first breaks into four species, *Lavendula stoechas, dentata, viridis* and *pedunculata;* the second into three, *L. officinalis* or true lavender; *L. latifolia,* the Dutch or broad-leaved one; and *L. lanata*.

Part of Lavender Classification (acc. D. A. Chaytor, 1937)

1. Stoechas
    - L. dentata
    - L. stoechas
    - L. viridis
    - L. pedunculata
2. Spica
    - L. officinalis=spica=vera
        - (a) L. angustifolia
        - (b) L. delphinensis
    - L. latifolia
    - L. lanata

The 8th edition of W. J. Bean's *Trees and Shrubs,* Vol II (1973) also deals with nomenclature of the lavenders and this work must now be reckoned the authority on the subject. The chief difference between Bean and Chaytor is that the former calls the true lavender *L. angustifolia*, synonymous with, instead of a sub-division of, *L. officinalis*.

L. lanata was discovered by Boissier in 1837, in Spain, where it grows in mountainous regions, especially on the Sierra Nevada. A charming, thickly woolly plant, one-and-a-half to two feet high, with bright, violet-scented flowers, and a strange strong fragrance reminiscent of menthol; it can usually be found growing as a rock plant in Cambridge Botanic Garden.

L. latifolia, is the spike lavender of the perfume industry, similar to true lavender in appearance although it flowers later; the flower spikes are more slender; the leaves broader; and a close examination reveals that the papery bracts attached to the stem below each cluster of flower heads are green, with one central vein. By comparison, the bracts in true lavender are brown and multi-veined. The major commercial source of spike is in Spain, where wild plants grow in the mountains at low altitudes. It is rarely grown in gardens; and where it is cultivated, it is more or less hyridized with *L. angustifolia*.

L. intermedia (known in the trade as lavandin) is a hybrid between true lavender (*L. angustifolia*) and spike (*L. latifolia*), and, as such, its properties come mid-way between those of its parents. It is a very hardy plant, long-lived, grows naturally at medium altitudes, and yields well. Although its oil does not compare for sweetness with that of true lavender, it is superior to the somewhat harsh, camphoraceous oil of spike.

One of the main complications in much of the early work on lavender is the confusion between true lavender and spike. Sometimes both were included under the name *L. spica*, sometimes it was used for one and sometimes for the other. In short, the word became completely ambiguous (see W. J. Bean for a detailed account).

At a meeting of the International Botanical Congress of the Pharmaceutical Society, held at Cambridge in 1930, it was decided that a name must be rejected if it becomes a permanent source of confusion; and soon after it was agreed that the correct name for true lavender would be *Lavendula officinalis* and that for spike, *Lavendula*

English lavender, detail from plate in 'Natural History of the Lavenders'
by Frederic de Gingins-La Sarraz.
(See Ref. 37, p.106).

French lavender (Lavendula stoechas). LEFT: A photograph of the growing plant. RIGHT: Detail from 'Natural History of the Lavenders' (See Ref. 37, p.106).

latifolia.[7] In theory, the principle still stands, though the first of these names has since been changed to *L. angustifolia*.

L. dentata is a native of arid situations in Spain, the Rock of Gibraltar, elsewhere in the Mediterranean area, and on some Atlantic islands. It is a bushy shrub of about one-and-a-half feet, and, like *L. stoechas*, the flowers are relatively closely packed, making a bulrush-shaped head of pale blue bracts and flowers at the top of the stem. It is easily distinguished by its dark green, regularly toothed leaves, and, like the rest of its section, it has sterile bracts on top of the flower head. But they are not as prominent as those of *L. stoechas* or *L. pedunculata*.

The characteristic features of *L. pedunculata* are, in fact, these violet terminal bracts. The plant is distributed throughout the Mediterranean region, though principally confined to Spain; and, like *L. stoechas,* it has square flower heads crowned with plumes of plump, purple bracts. Of the two, *L. stoechas* is marginally taller, and its native habitat extends eastward along the Mediterranean region to Greece. Both are on the tender side in this country, and both have a characteristic charm of their own.

A white form of *L. stoechas* was noted by Gingins-La Sarraz, and Messrs. Ellman and Sand found one with both white and purple bracts growing in the Eastern Pyrenees in 1925. The ordinary blue one was grown by Sir Frederick Stern in his garden at Highdown, Sussex; and it still grows there on the thin chalky soil.

Stained glass panel in Miss Lewis's museum. *(See description on page 89).*

3
Beginnings

Lavender has various uses: ornamental, herbal and cosmetic. It has been said to dispel melancholy weariness, and was once prescribed by Gerard for bathing on the temples to relieve "the panting and passion of the hart" and those that "swoune much",[8] besides a number of far less picturesque complaints! Why, Lear's Aunt Jobiska knew it was even efficacious for a toeless pobble:

> . . . his Aunt Jobiska made him drink
> Lavender water tinged with pink
> For she said, The World in general knows
> There's nothing so good for a Pobble's toes!

So, anyone who feels sceptical about some of the wilder claims made in lavender's name must remember that they have been made throughout history, and its history is a long one.

In common with many herbs and wild flowers, lavender has, from time to time, been credited with occult properties. Hecate, the Grecian goddess of infernal regions who presided over magic and enchantment, was patroness of witches and sorcerers. She and her daughters Medea and Circe knew all about herbs; and amongst those consecrated to this trio were the aromatic lavender, mint and corn feverfew. As a result, these herbs were persistently sought by witches who could appreciate their value and knew how to transform innocuous plants into deadly ones.[9]

Like other herbs, lavender has been used for healing from primitive times. Because many of the herbalists were medical men, medicine and botany were drawn together, mixed, of course, with a fair dollop of superstition and plant lore. Folk medicine grew up in all parts of the world, but the herbal history of lavender can be traced back to ancient Greece.

Dioscorides is credited as being the first to recognise the extensive use of medicines from all three of the natural kingdoms. He was well-travelled – possibly an army doctor – who spotted a vast number of interesting plants during his military forays. He was probably the first to mention French lavender (*L. stoechas*), calling it "An herb with slender twiggs having ye haire like Tyme, but yet longer leaved".[10] *Stoechas* grew near Gaul in the Islands of Stoechades (now known as the Islands of Hyères) from which it gained its name. This was the plant the ancient Romans chucked in their communal baths, the only lavender the Greeks knew.

Strangely, it seems as if the Greeks only used it medicinally. "Sharp in ye taste and somewhat bitterish, but ye decoction of it as the Hyssop is good for ye griefs in ye thorax", Dioscorides commented, attributing to it certain laxative and invigorating qualities, and recommending its use in a tea-like infusion for chest complaints.

The great doctor Galen (A.D. 129-199) added French lavender to the time-honoured antidotes, theriac and mithridate; and Nero's physician used it for anti-poison pills and in uterine disorders. All the earliest physicians practised autosuggestion, hypnotism and psychotherapy, though in somewhat different forms and under different names. Natural remedies were prescribed for natural disorders as well as for what we would now call psychosomatic ones.

The elder Pliny, for example, said French lavender mitigated "the pains of the bereft",[11] besides promoting menstruation; and "French nard" or lavender taken in wine was used for snake bites, stings, upset stomachs, liver, renal and gall disorders, jaundice and dropsy. He was the first to distinguish between French and true lavender (*L. spica* or *L. vera*), telling us that the second was used in stretching exotic perfumes used by the Romans.

We know that the Romans doused themselves in bath oils. They wore no underclothes or stockings; and bathing was rightly considered clean and salubrious. Public baths were an essential part of social life in the ancient city, becoming fashionable community centres for coteries of the day. Vast sums of money were spent on elaborate buildings where they bathed and oiled, strolled, and met one another, and the *Unctuarium,* or oil room, must have looked something like an old-fashioned chemist's shop, strung with jars and vases of perfume.

Of course, the Roman baths in this country ceased to be used soon after the Romans left; because the Church strongly disapproved of public bathing, on puritanical grounds. During the ensuing turmoil the baths fell into disrepair, and nettles sprouted between the mosaics.

We tend to assume that the Romans brought lavender to Britain although we have no definite evidence to support this. Even if they used it in this country, they may have imported it in the dry form; which brings us to a second conundrum: if they did grow it here, did it subsequently die out, or did it escape from cultivation, seeding itself in remote nooks and crannies into early medieval times?

The year 1568, the date of completion of the third part of Turner's herbal, has been quoted variously as the time lavender first appeared in British literature, and the time of its re-introduction from the Neapolitan hills. It has also been said that it was brought over by the Huguenots when they left France after the Revocation of the Edict of Nantes in 1685.

But the first really original work on gardening written in English mentioned "lavyndull". A treatise in verse by Master Jon Gardener, later called "A Feate of Gardening", gives decisive evidence of what was to be found actually growing in our gardens. The only copy in existence was written in about 1440; although the poem is thought to have survived from an earlier date, giving solid proof that, in common with a variety of plants believed to have been introduced in the 16th century, lavender was growing in England at least 100 years earlier.

4
Middle Ages

Since the beginning of the Christian Church in England, incense and fragrant flowers played an important part in ceremonies. You only need to smell the old lavender bush on an autumn bonfire to understand why the early Romans smouldered the stems to honour their gods; and lavender may have been the "incense" referred to in a charming riddle from the Anglo-Saxon Exeter Book:

> I am much sweeter than incense or the rose
> That so pleasantly in the earth's turf blows . . .[12]

In Spain and Portugal it was strewn on floors of churches and houses during festivals; and in Britain, especially on the Feast days of St. Barnabas and St. Paul, churches were decked with box, woodruff, lavender and roses; whilst officiating priests wore garlands of flowers in their hair.

It was monks who preserved knowledge of herbal lore and fostered the gentle art of gardening during the chaotic days of the Dark Ages. Since vegetables formed a large proportion of the monks' daily food, gardens were essential to monasteries. The plants cultivated in them were limited, but a plot was always set aside for herbs like lavender. We know that there was a continual flow of knowledge and plant material from Christian brothers on the Continent, and they were certainly familiar with the plant there.

A certain Abbess Hildegarde (1098-1180), from a Monastery in the diocese of Mainz, made some of the earliest medieval references to lavender, in her prolific writings.[13] Hildegarde, considered to be well versed in medicine and the natural sciences; described the plant as fierce, dry and strong-smelling, albeit without edible value. She realised its practical effect in getting rid of lice, and, like Langham, she prescribed it for clearing eyes. She had, into the bargain, a personal hunch that it could drive away evil spirits. I suppose a mixture of Flit, Optrex and pink pills for pale people, with a touch of the exorcist, might just about sum up her opinion!

Turning to lice, a fearful problem in the Middle Ages, one Anglo-Saxon herbal, which appeared in print shortly before Elizabeth I came to the throne, recommended lavender as a protection against "dirty filthy beasts"[14] (as well as for preserving chastity) "if the head is sprinkled with lavender water"; and William Langham wrote hopefully, "no lice will be in it [thy shirt] so long as it smelleth of it [lavender]."[4] It was still as a delousing measure, in 1874, that blotting paper was soaked in the oil before being applied to the heads of school children in Provence; whilst Mrs. Grieve, a modern herbalist, writes that "it is said on good authority that the lions and tigers in our zoological gardens are powerfully affected by the scent of lavender water, and will become docile under its influence".[15] Perhaps some lions and tigers are more sensitive than others.

Lavender was listed with plants of medical virtue in the *Meddygon Myddfai,* a famous collection of recipes for various diseases, injuries, prognostics and charms, written by the celebrated Welsh physicians of Myddfai in Caermarthen about the middle of the 13th century. It was also mentioned in *Hafod,* a Welsh text dealing with medical matters, c.1400; and is actually incorporated in some of the cures handed down from the skimpy literature of the period. This one for making *aqua water* is irresistible:

> Take sauge and fynel-rotes and persely-rotes and rosemarynę and tyme and lauendre of euerech lyche moche and wasche hem and drye hem after and wenne they ben drye, grynde hem a lytel in a morter and strawe ther-on a lytel salte, and putte hyt in the body of the styllatorye and helde [pour] there-on wyne, reed other whyzte [red or white] thene putte hyt in a potte fulle of asckes [ashes] ouer the forney [furnace] and makę so softe fuyre ther-under that wen the styllatory by-gin to dropp, loke that hyt dropp no fastur than thou myste seyze [say] on, two, tre, betwene the droppys.
>
> And so do stylle hyt al to-gedre; thenne take thye water that is distillyd, and distyllet [distil it] azen zyf [as] you wolte and use that of euerech day a lytel sponeful fastyng.

Herbert Schöffler gives two recipes in his work on medieval medical literature. In the first, lavender is ground with ambrose and sage, fried well in butter and drawn through a cloth to make an ointment for "sore bak and sore reynes [kidneys] and sore sydes".

In the second, it is mixed with watercress, rue, sage, juniper and oximel (*Ononis spinosa*) for palsy; or powdered and placed beneath the tongue to "Make a man speke that have lost hys speche".[16]

Some medieval brews sound so far-fetched that you could be excused for wondering whether there was even a vestige of science behind them. For metheglin, a special mead also used as a medicinal liquor, lavender roots were combined with those of fennel, parsley, elecampane, radish, wormwood, valerian, herb-Robert and so on; then they were all boiled in river water, cooled and sieved, before reboiling, skimming and bottling. As the resulting liquid had to stand for six months before being used, the whole thing seems to have been a major operation.

For "gout that is cold" the procedure was less murky: "Take cowslip, sage, lavender, rednettle and primrose, and seethe them in water and wash the place with the water, and it shall be helped".[17]

A physic garden where herbs were grown and their oils distilled. Note the stills (top left and bottom right). The illustration is from the early 16th century (see acknowledgements).

5
Herb Gardens

Here's flowers for you;
Hot lavender, mints, savory, marjoram;
The marigold, that goes to bed wi' the sun,
And with him rises weeping; these are flowers
Of middle summer . . . *(A Winter's Tale)*

Lavender has first place in Perdita's floral catalogue, although this is the only time it crops up in Shakespeare's vast works. The adjective "hot" is a bit of a mystery; I am inclined to agree with Alan Dent in his *World of Shakespeare: Plants,* that it means simply pungent. The herbalist, William Turner, called it hot in *Names of Herbs,* whilst Gerard called it "hot and dry".

Spenser only mentions it once, too: "the lavender still grey";[18] and Bacon, not at all, in his inventory of sweet-smelling plants.

Does this imply that lavender was rarely grown in Shakespeare's time? I think not. The golden age of herbs and herb gardens in Britain stretched roughly from the late fifteenth until the mid-seventeenth century. When the wars of the barons and knights ended, and relative peace reigned; the lords, no longer rivals in the battlefield, set out to vie with each other in the size and grandeur of their estates.

Herb gardens reached a peak under Elizabeth, after which they gradually declined in favour of more general flower gardens where plants were grown for their own sake, rather than their usefulness. True herbals gave way to flower books packed with information on new species, which were pouring into the country from abroad; and herbs no longer had to be grown on the same scale, since they could be bought in markets or in the streets.

But the practice of reserving a special area for herbs was continued so long as their importance merited cultivation in places where they would always be at hand. In fact, herbs prospered in kitchen and in pleasure gardens. Thomas Hill, author of an early Elizabethan gardening book called *The Gardener's Labyrinth,* recommended separate beds for individual species: "It shall be right profitable to level a bed, only for Sage, another for Isop, the like for Tyme, another for Marjorum, a bed for Lavender . . ." and so on. The rest of the garden also ought to have its quota of "physick herbs", he thought, for in addition to being "a delectable sight", they "yield a commodity to our bodies, in curing sundry griefs". Mazes and knot beds were the height of fashion until the Restoration: every gardening book devoted pages to their design and care.

On cultivation, Hill suggested that herbs like "Sticas or Lavender gentle which is woundrous sweet, both leaf and flowers", should be sown in the spring, from March to April, "for they speedier come forward, then bestowed in the month of February". Seeds sown during the moon's first quarter, in sunny, warm places and after rain, rooted much faster, he said, while slips were best set about Michaelmas. The flowers should be dried on plates in August, in full sun. "Gather them as you dry them, when you see the morning fair and hot, and the hearbes dry".

William Lawson thought that, on the whole, flowers were better kept away from onions and parsnips; although lavender was "sufficiently comely and durable" for squares and knots, and indispensable in the kitchen garden "where aesthetics ought not to be altogether neglected".[19]

The space devoted to herbs and herb gardening in today's women's magazines tends to suggest that they are often considered a woman's realm; and, largely because she has traditionally taken first place as cook, this has long been the case. It was a division of labour noted by Sir Anthony Fitzherbert in his *Book of Husbandry* (1522). Whilst the husband ploughed the land, planted pulses and fed cattle, his wife was tending herbs. "The beginnynge of Marche or a lyttell afore, is tyme for a wyfe to make her garden and, to gette as many good sedes and herbes as she canne, and specially suche as be good for the potte, and to eate: and as ofte as nede shall requyre, it must be weded, for els the wedes wyl ouergrowe the herbes"; while the Elizabethan, Thomas Tusser,[20] listed "lavender of all sorts" under "herbs and branches and flowers for windows and pots", as well as under "herbs for strewing"– both wifely duties; and William Lawson dedicated an entire work to women.

"Garden herbs are innumerable, yet these are common, and sufficient for our country house-wives". He suggested that spike lavender bushes be discarded after seven or eight years; the white kind even sooner, and he called the latter "most comfortable for smelling except rotes (the rootes)", and "good for bees". Lawson also noted that lavender, kept dry, is as strong after a year as when it is gathered.[19]

Of course, herbs were once used in cooking far more lavishly than they are today; one of the reasons being that, in medieval times, and even later, fresh meat was rare after September. It was at best coarse, at worst tough and unpalatable; so spices were necessary to disguise the unpleasant flavour of imperfectly preserved flesh. Most people are familiar with mint in cooking, but how many have used powdered lavender? Apparently, Queen Elizabeth I enjoyed a conserve of lavender, which was always on her table.[21]

For the Tudor or Stuart housewife, with a family to clothe and feed, and a house to keep clean, perhaps herb gardening lent a spot of relief to relentless routine. Gervaise Markham, at any rate, expected a high standard from women: "Let your English housewife be a godly, constant, and religious woman, learning from the worthy preacher, and her husband, those good examples which she shall with all careful diligence see exercised amongst her servants."[22] She should also be "temperate, of great modesty and temperence, inwardly as well as outwardly". Her garments "comely and strong altogether without toyish garnishes or the gloss of light colours", and her character peerless, "chaste thoughts, stout courage, patient, untired, watchful, diligent, witty, pleasant, constant in friendship, full of good neighbourhood, wise in discourse, sharp and quick in her affairs, comfortable in her councils, and generally skilful in the worthy knowledges which do belong to her vocation".

A tall order, one might imagine, for any aspiring Mistress Markham.

6
Herbalists

William Turner was a pioneer among British herbalists, a versatile physician, botanist and divine, often called the Father of English botany. He wrote a lusty, independent herbal dedicated to Queen Elizabeth I. He did not add to the uses of French lavender, beyond noting that it is good mixed with treacles and preserves, but true lavender he recommended for all diseases of the brain that "come of a cold cause". Mixing with cinnamon, cloves, mace and the leaves of rosemary, he said, strengthened its healing power; whilst a water distillation cured certain types of headache: "And so it helpeth the dulness of the head".[2]

Time has turned the next great English herbalist into one of the best known, though to some extent undeservedly; John Gerard, barber-surgeon, got hold of a translation of a French botanist's final work (Dodoen's *Peptades*), altered the arrangement, and used it freely as though it were his own, even denying that he had seen the original. If his unmistakable prose has a characteristic quaintness, many of his recipes were either unscientific or inaccurate. Nonetheless, modern thought is inclined to credit Gerard with a large part of the herbal which bears his name; and this alone is a monumental achievement.[23]

Gerard makes no separate entry for French lavender, but true lavender (which he calls spike), he pronounced effective against catalepsy, migraine and fainting, cautioning that in certain cases, however, it could be dangerous. In fact, foolishly administered, it would "often times bring death itself". Elsewhere he suggested powdered petals mixed with spice and distilled water or oil for palsy. He particularly recommended a bean-sized portion of lavender conserve, taken at breakfast, as a strong dose.[8]

Not only in herbals, but in almost every contemporary practical work on gardening, the "vertues and physic helps" of flowers were listed, in the belief that "many herbes and flowers with their fragrant sweet smells doe comfort and as it were revive the spirits". The words belong to John Parkinson, apothecary to King James and renowned for his writings. On his authority, lavender was used "both to accomplish this garden, and to please your senses, by placing them in your nosegayes, or elsewhere as you list". He said it was used for scenting linen, clothes, gloves and leather; and that the dried flowers would "comfort and dry up the moisture of a cold braine". Small lavender, or spike, was virtually only used externally, whilst French lavender was "somewhat sweete, but nothing compared with lavender".

Parkinson's *Paradisus* was dedicated to Queen Henrietta Maria, winsome young wife of Charles I, who grew white lavender (also, apparently, a favourite of Nell Gwyn's) at the Manor of Wimbledon.[24]

In the 17th century, England was overrun with what has been called "an infection of astrological botany". Every disease, according to the exponents, was caused by a planet; and, since certain herbs also came under the dominion of the celestial bodies, one way of curing an ailment was by balancing it with a herb from an opposite planet. This meant that diseases produced by Jupiter, for example, would be cured by the herbs of Mercury. Alternatively, you could use sympathetic herbs belonging to the same planet that caused the disease.

Setting up as astrologer and physician, Nicholas Culpeper waxed eloquent in the game of speculation and conjecture. A past-master at spreading his gospel, he gathered a large audience in London's East End; and his herbal ran into edition after edition, despite its shaky validity. In spite of everything, however, he was a good physician and did much to help the poor.

Lavender, he decided, "is owned by Mercury, and carries its effects very potently". A decoction made from a mixture of its flowers with horehound, fennel, asparagus root and a little cinnamon, he approved for fainting and giddiness; and a gargle made from a brew of the same strange mixture for toothache. Two spoonfuls of the distilled water helped cure loss of voice, and "the tremblings and passions of the heart"; which sounds as if it might be interpreted physically or psychologically, since he recommended a few drops of the "fierce and piercing" oil of spike for "inward or outward griefs".[25]

Despite Culpeper and a few other practitioners, Parkinson was really the last in the line of great English herbalists, leaving us with roughly another two and a half centuries before the onset of modern medicine, and a little more than three to the present day. During this period, reports of lavender in medicine reached a peak in William Salmon's herbal (1710), then dwindled gradually in the light of chemical science.

Salmon said that lavender is "Abstersive, Aperitive, Astringent, Discursive, Dieuretick and Incisive, . . . Cephalick, Neurotick, Stomatick, Cordial, Nephretick and Hysterick. It is Alexipharmick, Analeptick, and Antiparatitick, being of very subtil and thin parts". He approved it fairly sweepingly for convulsions, epilepsies, palsies, tremblings, vertigos, lethargies, swoonings, hysteric fits and other diseases of the head, brain, nerves and womb, in twelve different preparations.

To Pliny's recommendation that it be used for snake-bites, Salmon added the bites of mad dogs and "other venemous creatures". It could be applied externally as a poultice or internally as a "spiritous tincture" of the dried leaves and seeds. The latter was taken for hysterical fits, however violent and long-standing.

Two hundred and fifty years later, a tisane; or even a spray of lavender worn under a hat, "as harvesters themselves apply it", was being commended for nervous headache according to Mrs. C. F. Leyel. Mrs. M. Grieve, the author of *A Modern Herbal* affirmed that in some cases of mental depression and delusions, oil of lavender is genuinely helpful, whilst a few drops rubbed on the temple relieve headache; and compound tincture, sold by the name of lavender drops, was, at the time of writing, still a common remedy for faintness.

In 1931, Mrs. Grieve notes, the oil was used for swabbing wounds, sores, varicose ulcers, burns and scalds and for other antiseptic and surgical purposes. She asserted that in France it was quite general for households to keep a bottle of the essence as a homely remedy for bites, bruises and trivial aches and pains, and she revealed that it was also used in veterinary practice for killing lice, worms and other parasites.

Of course, its use as an insect repellent had been recognised since Hildegarde's time. This was pointed out in *The Chemist and Druggist* in 1910, when the oil was being used for flea-bites, said to be largely responsible for an epidemic of spotted fever in the Midland counties. Lavender was used, it said; though doubtless there were less expensive oils that would do just as well.

In 1972, the Australian, Dorothy Hall, reaffirmed its potency, particularly advocating a few drops rubbed on the skin before bushwalking or barbecuing, to repel midges, mosquitoes and flies. In *The Book of Herbs* [27] she says that herbs gathered from thousands of British gardens were used extensively in field hospitals during the Second World War for combatting infections and purifying surgical dressings. It was even used for cleaning wards when disinfectant was scarce.

However, according to tests carried out in the United States on the properties of lavender oil, it is only slightly anti-bacterial, although it is more markedly anti-fungal. This may give some foundation for its clinical use.

An even more recent account by Professor Hans Flück[28] indicates that the properties claimed for the plants today are much the same as in earliest history. Nor have the needs of man changed so very much. A recent issue of *Harpers and Queen* magazine[29] promotes an age-old Indian system of health care, based on the idea that regular massage with potent plant essences affects both mind and body. Lavender essence is given high priority for its anti-depressant, sedative and diaphorectic qualities.

A medieval physic garden from a book published in 1557 (see acknowledgements). It shows the preparation and distillation of herbs (left), doctors conferring (right) and a patient in a "hospital" (top left).

7
Distilling

The Egyptians used stills to extract essential oil from the brilliantly tinted, heavily scented flowers that grew around them in such profusion, and their secret found its way across the blue waters of the Mediterranean to Italy and Greece.

Essential oils lend characteristic odours to bark, buds, leaves, roots, flowers or whatever parts contain them. They are made and stored in minute glands in the tissue of fragrant plants. In lavender, these glands lie chiefly in two parallel straight lines on the upper lip of the petals and between the hairy ribs of the calyx.

The oil is yellowish, with a strong taste and smell; practically insoluble in water, though easily soluble in alcohol. It is volatile and so can easily be distilled. Extraction takes place by heating water and passing steam through the plants, turning the oil into gas. The mixture of steam and gas passes into a condenser, where the gaseous lavender oil, being more volatile, liquifies first. The distillate soon separates out into two layers with the oil on top. Until 1906 the universal method of extracting lavender oil was by water distillation; the flowers being literally immersed in water, which was then heated, and the oil carried over in a condenser with water vapour. This is still practised in Southern France, although it was superseded in this country by the more rapid steam distillation process. Both work on the same principle, the difference lying in the way in which the plants are treated at the beginning of the process.

Prior to the 16th century there are few records of distillation in England; then, quite suddenly, it became part of the routine in all large houses. Balm, sage, marigold and tansy were distilled in plenty, and every garden had its rosemary and lavender bushes. The earliest known recipe for lavender water, dated 1615, gives directions for distilling the flowers with canella, wallflower, gallingall and the grains of paradise – whatever they may be. Modern lavender water is made by diluting the essential oil with alcohol and blending it with other ingredients; but in early recipes it was simply distilled in water.

Today, lavender oil is in regular demand, chiefly for perfumery and the manufacture of toilet soaps. The quality of the oil is often considered a direct correlative of its ester content. But it is the odour which is of real value as a criterion of quality and suitability.

There are two main types of the oil recognised on the market: the true one, from varieties of *L. officinalis;* and lavandin, which is made from hybrids between *L. officinalis* and *L. latifolia* (spike).

Oil from pure spike, known in commerce as oil of spic, or aspic, has an acrid taste and a strong, disagreeable, turpentine-like smell. It is made largely in Spain and used traditionally in veterinary medicine, or for certain quick-drying varnishes.

We have already remarked that the production of essential oils did not become general until the second half of the sixteenth century. In 1500 and 1507, two volumes of Hieronymous Brunschling's famous book on distillation, *Liber De Arte Distallandi*, appeared at Strasbourg. The author, a physician, mentions four essential oils: oil of turpentine (known since antiquity), oil of juniper, and oils of rosemary and spike.

But, although medieval days were basically anti-indulgent, and women who used scent and make-up were denounced by the Church as bedizened whores, young squires and nobles took regular, scented baths, and when the first wisps of Renaissance began to percolate through from the Continent, the lavish Italian fashion for scent caught on in England. In the seventeenth century, distilling became as popular a hobby as wine-making is today. "Sweet waters" were presented at Christmas or birthdays, whilst court ladies whiled away the hours in their still-rooms amongst bunches of drying flowers and herbs, rows of mortars and pestles, fixatives and exotic spices, arranged on long wooden shelves. Still-room secrets were sometimes recorded in special books for the next generation, sometimes simply passed down orally from the mistress of the house.

From fields they gathered rushes and sweet-smelling grasses to throw on their floors; in the garden they cultivated medicinal and aromatic herbs; and, in their still-rooms, they powdered, mixed and stilled, transforming the summer's harvest into moth-bags, sweet-waters, pot-pourris, sweet-bags, pomanders, wash-balls, sachets, herb-pillows, tussie-mussies, vinegars and teas.

When James I visited the Bodleian Library at Oxford, the floors had been rubbed with fresh rosemary, and the furniture with oil of lavender; a precursor, I suppose, of modern waxing and polishing. Mixtures of woodruff and lavender were hung in homes and churches as air sweeteners in Elizabethan England and America; and Parkinson described the delightful custom of tieing small bundles of costemary, lavender and rosemary to "lie upon the tops of beds"[24] (I think he meant on pillows) "there to impart their fragrance".

Tussie-mussies, romantic little nosegays of herbs and flowers, may well have originated as medicinal bouquets carried to combat the unpleasant oduor of bad sanitation, and fight away plague germs. Also used during plague epidemics, when the price of appropriate herbs would rocket, were herbal vinegars, which could either be soaked into sponges for sniffing, bathed on the temples, or enclosed in tiny compartments of doctors' and gentlemen's walking sticks.

You do not see so much variety in walking sticks today. How well I remember the selection of my great grandfather, which hung around the house when I was a child. There was a hefty twirl of ivory finished with a round blob, and there were elegant, slender, black canes, silver hooked and tipped, which we used for pulling down blackberries. But the one we prized the most held a minute glass container and tiny drinking cup on either side of its handle.

On the whole, herbal vinegars were not intended to be taken internally, despite the fact that they were made in much the same way as the vinegars now used for salad dressings. During epidemics, men and women sprinkled them over linens and bedclothes, and they were used to scent baths and make them antiseptic. I have even heard that lavender vinegar was applied by women to colour their cheeks as an innocent form of rouge.

Pot-pourris are old favourites. The word actually means rotted pot, the pot being a jar where the pourri is rotted or fermented. There are any number of variations, but recipes generally include roses and lavender, since they retain their fragrance so well when they are dried.

One cherished by Charles II's wife, Catherine of Braganza, was made of three parts rose petals, with dried rosemary, thyme and lavender, the powdered skin of an orange and some cloves. Another recipe, popular in the sixteenth century, included lavender, orris-root, vanilla bean, clover, allspice, bergamot, rose petals and cinnamon.[30]

Mrs. C. W. Earle, an accomplished Victorian housewife and writer, placed scented bags around her furniture: "On the backs of my armchairs are thin Liberty silk oblong bags, like miniature saddle-bags, filled with dried Lavender, Sweet Verbena and Sweet Geranium leaves. This mixture is much more fragrant than the Lavender alone. The visitor who leans back in his chair, wonders from where the sweet scent comes."[31]

Even today, children sew them in school; and, since the days of Elizabeth I, rich and poor have made sweet-bags for their houses: to place in drawers, among linen, or in the bookshelves – anywhere.

The most common herbs for sweet-bags and pads, along with rose petals, rosemary and marjoram, were basil, thyme, coriander, anise, caraway and lavender. The spices and fixatives were those for pot-pourris–cinnamon, clove, nutmeg, orris-root powder, benjamin; but the composition was headier, since the scent had to permeate fabric.

> Velvet gown and dainty fur
> Should be laid in lavender,
> For its sweetness drives away
> Fretting moths of silver-grey[32]

People realised from earliest times that moths were deterred by lavender. An exception was Sir Matthew Hale, who connected the two but jumped to false conclusions. We may forgive him: he was an historian and lawyer, not a gardener. "The seeds of lavender, kept a little warm and moist", he said, "will turn into moths".[33]

Sachets and sweet-bags were filled with lavender to guard against moths, and a special way of impregnating clothes, by sprinkling them before ironing, was advocated in a slightly muddled seventeenth century recipe:

> To make a special sweet water to perfume clothes in the folding being washed. Take a quart of Damaske-Rose Water and put it into a glasse, put unto it a handful of Lavender Flowers, two ounces of Orris, a dram of Muske, the weight of four pence of Amber-greece, as much Civet, foure drops of Oyle of Clove, stop this close, and set it in the Sunne a fortnight; put one spoonfull of this Water into a bason of common water and put it into a glasse and so sprinkle your clothes therewith in your folding . . .[34]

I cannot resist another recipe for "Sweet water for Linens" from *Bulleins Bulwarke* (1562), because there is something in the literature of the period that makes pure poetry out of quite ordinary household hints.

Three pounds of Rose water, cloves, cinnamon, Sauders [sandalwood], 2 handfull of the flowers of Lavender, lette it stand a moneth to still in the sonne, well closed in a glasse; Then destill it in Balneo Marial [bain-marie]. It is marvellous pleasant in savour, a water of wondrous sweetness, for the bedde, whereby the whole place, shall have a most pleasaunt scent.[30]

Harvesting lavender – child's play?

8
Trade

Today it would be fair to say that all major perfumers use lavender in one way or another, and most fern types of scent contain its oil; but perhaps Yardley is the first name we associate with lavender water.

Way back in the reign of the autocratic Charles I, the family first entered the beauty business when a young Yardley paid the monarch a large, and doubtless welcome, sum of money for a concession to manufacture soap for the entire city of London. Alas! full details of his enterprising activities were lost in the Great Fire. Just one detail is certain: he used lavender perfume in his soap.

It is not until the eighteenth century that we hear of the Yardleys again in business circles; this time occasioned by a certain William making the acquaintance of Mr. Beedal, a prosperous steel-buckle maker. Yardley profited from this friendship. Before long, he had learned enough about commerce to set up his own sword, spur and buckle business, and acquired sufficient worldly wisdom to marry Beedal's widow just six weeks after her husband's death. Of the seven children from this match, one daughter, Hermina, married William Cleaver, heir to a big soap and perfumery business. It was she who helped found the House of Yardley in 1770.

Across the Channel perfumery was having a chequered career. That very year, an Act of the French Parliament (as quoted in Edward Sagarin's *The Science and Art of Perfumery,* McGraw Hill, 1945) read:

> That all women, of whatever age, rank, profession, or degree, whether virgins, maids or widows, that shall from and after such Act, impose upon, seduce and betray into matrimony any of His Majesty's subjects by the scents, paints, cosmetic washes, artificial teeth, false hair, Spanish wool, iron stays, hoops, high heeled shoes, bolstered hips, shall incur the penalty of the law in force against witch craft and like misdemeanours and that the marriage, upon conviction, shall stand null and void.

How many such annulments, we might wonder, were granted by male judges from their benches, their long, smug false tresses hanging from bald heads?

In Britain a revolution in dress was under way. Men's wigs were going out of fashion, much to the alarm of the makers, who actually petitioned the King for redress. The newly wigless now wore only a dressing of bear's grease, scented with lavender, on their hair. Perfumes, however, were simple and unsophisticated, mainly produced by macerating flowers and distilling them in water.

By the end of the eighteenth century, the use of lavender perfume was on the wane. No longer did people drench themselves in scent, nor were lavender flowers scattered over floors, or stems burnt in rooms. Gone were the days of perfumed baths, and rings that ejected fragrant spurts. Just as manufacturing perfumers were beginning to produce the greatest variety of attractive scents the world has ever known, European women could think of little use for them beyond placing two drops behind their ears!

It was also an age of uncertain fortunes: extravagance was fashionable, riches soon lost and won. Our friend William Cleaver ran short of money, and his father-in-law came to the rescue. From this point the Yardley business, boosted by the addition of the Cleaver enterprise, took on a new lease of life.

Early in the twentieth century, a London centre was opened at 8 New Bond Street, and an extensive advertising scheme was designed to foster the demand for lavender products. During the '30s, cosmetics again became respectable. If society beauties featured in glossy magazines had acquired their charms by using perfume, surely no mother could refuse her daughter the chance of becoming irresistible overnight.

The Yardley shop, 8 New Bond Street, opened in 1910.

9
Victorian Whimsy

Queen Victoria is said to have had great faith in lavender. "The Royal residences are strongly impregnated with the refreshing odour of this old-fashioned flower and Her Majesty has it direct from a lady who distils it herself".[35]

As a practical person, the Queen would probably have used it as a domestic disinfectant. But so often was it used in the Victorian home: sewn into scented bags or washed on floors and furniture, that it became almost a symbol of the times. From contemporary rhymes and expressions we have some idea of its appeal.

"Laid up in lavender" arises from the old habit of sprinkling the herb among linen to preserve it from moths, and to "do something up in lavender" means to treat it with special care. "What woman, however old, has not the bridal-favours and raiment stowed away, and packed in lavender, in the inmost cupboards of her heart?" asks Thackeray in *The Virginians*.

Several late-Victorian writers link lavender with distrust, because it was once thought that the asp, a dangerous species of viper, lived in the plant.[36] This meant that it had to be approached with extreme caution. In Tuscany it was used to counteract the effects of the Evil Eye on small children, and, strangest of all, Kabyl women apparently believed it protected them from marital cruelty.[9]

But it seems as if lavender acquired a slightly precious quality, pre-dating blue jeans and women's liberation. Its medicinal virtues have been reputed particularly effective for the blue-blooded: "It is also an agreeable and available remedy for steaming, and for headaches, and other slight maladies to which persons of gentle breeding are subjected", says de Gingins-Lassaraz, Swiss author of a learned monograph on the plant.[37]

Equally mawkish sentiments were expressed in an article in the *Mitcham News* of 20th July, 1934:

It seems to go back through the mists of years, dim sweet forms arrayed in flowered muslin crinolines and poke bonnets, gathering the fragrant herb to make their sweet pots and dainty sachets.

There are certain types of women whose individuality seems to be better expressed by the use of lavender water than by any other scent. Gracious, gentlewomen, with a love of all that is fair, harmonious and beautiful in life and thought, rather than the sophisticated, dashing type.

Lavender and old ladies were almost fated to go together. William Shenstone's schoolmistress, for instance, was particularly fond of it:

> And lavender, whose spikes of azure bloom
> Shall be, ere-while, in arid bundles bound
> To lurk amidst the labours of her loom,
> And crown her 'kerchiefs with mickle rare perfume.[38]

Square or heart-shaped bags of pink or lilac net, tied with velvet ribbons, were traditional gifts with which a maiden aunt honoured the birthdays of her young nieces; and, in *Cranford,* Mrs. Gaskell speaks of "little bunches of lavender flowers sent to strew the drawers of some town-dweller, or to burn in the room of some invalid".

It was frequently mentioned in honeyed Victorian verse; but perhaps a poem in a school magazine, written by a youngster growing up in Hitchin during its lavender days, gives as good an idea as any of its slightly prim associations:

> Lavender is the colour of innumerable
> Grannies pressed between the leaves
> Of an old book.
> Lavender is a gentle pillar of the
> Establishment,
> Lavender does not approve of permissiveness
> Lavender is the fragrance of half
> Forgotten things . . .[39]

I think that all the 'old-maidish' connections are a little surprising. After all, the scent might well be described as 'invigorating', 'thrusting' or 'fresh and spicy'; surely more appropriate to the young?

John Parkinson said the smell on a handkerchief would "pierce the senses in a most refreshing manner".[24] It was, moreover, one of the flowers strewn on the benches of the Old Bailey to distract the courts from the smell of nearby Newgate Prison. Judges still commemorate the custom when they carry a sprig in their buttonholes at the beginning of the season.

"Lavender's Blue", the popular children's nursery rhyme, had its source in a playful love song printed some time between 1672 and 1685 on a black-letter broadside:

> Lavender's green, diddle diddle,
> Lavender's blue
> You must love me, diddle, diddle,

Cause I love you,
I heard one say, diddle diddle,
Since I came hither
That you and I, diddle diddle,
Must lie together.[40]

There were ten verses in all, and over a century later it emerged in abbreviated form as a nursery rhyme.

Lavender blue and Rosemary green,
When I am king you shall be queen;
Call up my maids at four o'clock,
Some to the wheel and some to the rock;
Some to make hay and some to shear corn,
And you and I will keep the bed warm.[41]

My guess is that many a Victorian toddler was bounced on its mother's knee to the less usual version:

Lavender's blue, little finger, rosemary's green,
When I am king, little finger, you shall be queen,
Who told you so, Thimby? Thimby, who told you so?
'Twas my own heart, little finger, that told me so.[42]

At any rate, the rhymes were remembered almost solely in the nursery until a dance version, popularly called "The dilly dilly song", swept America and Britain. Several similar versions occur in eighteenth and nineteenth century children's literature; in *Gammer Gurton's Garland* the words are "Roses are red, diddle, diddle, Lavender's blue: If you will have me, diddle, diddle, I will have you"[43]

Finally, a rhyme for St. Valentine's Day which went:

Lilies are white Rosemary's green;
When you are king I will be queen.
Roses are red Lavender's blue;
If you will have me, I will have you.[44]

In all of them, strong rhythms with lilting tunes, lavender is used for the rollicking bounce of its three syllables and because of its colour: against rose's red, its vivid blue.

One more association inevitably bound with lavender is its Englishness. Never mind that the bulk of raw materials now comes from France; throughout the country there are still place-names harking back to the old lavender-growing days: Lavender Ways, Lavender Roads, Lavender Closes, Lavender Farms and, perhaps best known of all, Battersea's Lavender Hill. The imagination is almost unequal to the task of trying to reconstruct this area as it was a century ago: Lavender Hill as a place of green meadows and

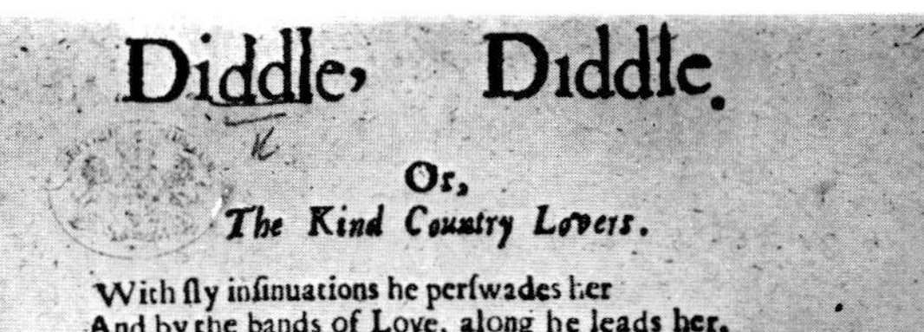

Diddle, Diddle.

Or,
The Kind Country Lovers.

With ſly inſinuations he perſwades her
And by the bands of Love, along he leads her.
Relating pleaſant ſtories for to bind her
And all to make her unto him prove kinder.
And ſo in Love at laſt they live together
With pleaſant dayes enjoying one another.

Tune of Lavender green, &c.
With Allowance, Ro. L'Eſtrange.

a. Headpiece for the song 'Lavenders green, Diddle diddle', c. 1680. British Museum

b. Cover for the song 'Lavender blue dilly, dilly', 1948

A page from the 'Oxford Book of Nursery Rhymes' (see Ref. 40, p. 107).

nearby trout streams! As a matter of interest, there is no written evidence to suggest that lavender was grown there, but a number of elderly residents remember it on local holdings around the turn of the century. "Lavender Sweep is drowned in Wandsworth",[45] writes John Betjeman, who manages to save the most dismal contemporary scene with a wisp of nostalgia.

Piesse and Lubin, described in *The Chemist and Druggist* in 1874 as "the celebrated perfumers of Bond Street" owned a distillery and herb gardens near Mitcham Common. Septimus Piesse wrote:

> In each bright drop there is a spell
> 'Tis from the soil we love so well
> From English gardens mown.[46]

A certain Caryl Battersby conjured up an aura of whimsical euphoria in a series of rhyming couplets quoted in an early twentieth century brochure on lavender:

> Ladies fair, I bring to you
> Lavender with spikes of blue;
> Sweeter plant was never found
> Growing on our English ground.[32]

10
Mitcham Physic Gardens

"Although the cultivation of medicinal plants is carried on in various parts of England, yet more land is employed in this way in Surrey than in any other county". Thus reads the *Pharmaceutical Journal and Transactions* of 1850-51.

The arms of the former Borough of Mitcham carry two sprigs of lavender on a gold backcloth; the Borough colours were green and mauve; and the cloth bands worn with the Mayor's official badges were also mauve. Mitcham also boasts the 'Gardeners' Arms', the 'Mitcham Mint', Lavender Avenue, Camomile Avenue, Rose Avenue and Lavender Grove to remind us of the old physic gardens. But it was a very different Mitcham that grew herbs in its midst; one scarcely recognisable in the bustling suburb of today.

One of the earliest references to lavender in the area appears in the records of Merton Priory in 1301 (published, with a commentary, in 1898); when, partly from the sale of seed, and partly from tenants' donations, the neighbourhood raised £50 to lend to King Edward I. Amongst a list of items sold comes "Spikings—44 quarters" which is explained as "spiking, spike-lavender" with the comment that many acres of land in the immediate neighbourhood of Merton and Mitcham were still, in 1898, devoted to growing lavender and garden herbs.[47]

By 1700, the land was predominantly arable, and already the neighbourhood had acquired a reputation for being salubrious and select. "Noted for good air and choice company", the author of Murray's *Handbook of Surrey*[48] observed. The parish was bounded then, as it is today, by Streatham on the north-east, Beddington, Carshalton and Croydon on the south, Morden on the south-west and Merton on the west. It was, of course, conveniently near to London, a mere nine miles from Westminster Bridge; and communications improved when the London-Dorking Turnpike opened in mid-century; followed soon after by a stage-coach service from Mitcham via Tooting running four times daily. Not surprisingly, the area attracted London bankers and wealthy businessmen, who soon began building the residences that brought about Mitcham's transformation from an agricultural community to a popular suburb.

It was not, however, all rustic and provincial. Indeed, there appears to have been industry of some kind carried on along the banks of the meandering Wandle as far back as records exist. By the nineteenth century the Wandle was nominated the hardest worked river of its size in the realm.[49] Some of its low-lying water meadows were trenched for spreading out coarse cotton cloth from the calico-printing and bleaching industries; the quality of its water suited the silk-dyeing industry; and subsidiary work, such as the manufacture of felt and beaver hats, was introduced by the Huguenots of Merton Abbey. All day long the water-wheels turned. Here and there the river banks became disfigured with industry, and the idyllic herb fields were fringed with what James Thorne referred to as "far from fragrant factories".[50]

Gentlemen built elegant mansions with sweeping estates, whilst the labouring classes dwelt in the unsophisticated clapboard cottages which became such a feature of the Wandle area and still exist, dotted about the district. The importance of agriculture is clear from frequent references to farmer, husbandman, shepherd, and cowkeeper in the Mitcham Settlement Examinations (1784-1814).[51] Many examinants are described as labourers; and one "singlewoman", Elizabeth Weller, hired herself as a yearly servant to a Jo. Simmons of Mitcham, physic gardener, at wages of six guineas. That she served one year, and received these wages, is signed with a cross because she was illiterate. But if the workers' hours were long and the pay frugal, perhaps they were to some extent compensated by the spectrum of local colour described by Hassell: "Blue from the ripe lavender; red and brown from the herbs; rich dark yellow from the wheat; pale yellow and greens of various casts, from ripe and unripe barley and oats; purple from seed clovers; and deep brown from the fallow fields."[52]

There were, even in these early days, physicians or qualified doctors and they made full use of medicaments derived from herbs; but their services were primarily enjoyed by the rich, because their fees were high. So the apothecary, who made the physician's prescriptions, and stocked all kinds of herbal concoctions, to some extent took on the rôle of general practitioner and druggist in poor parts of a town. Many of them obtained large practices and considerable reputations. The Society of Apothecaries' Physic Garden had been set up on Chelsea Embankment; but its rôles, basically educational and scientific, made it quite inadequate for the expanding herb requirement, which was met by trundling supplies into the city daily from Mitcham's farmers turned herb growers.

Towards the end of the eighteenth century, Mitcham had 540 houses. Of 883 families, 296 were occupied chiefly in agriculture, 460 in trade, handicrafts or manufactures; and about 250 acres were occupied by physic gardeners in the cultivation of herbs such as lavender, wormwood, chamomile, aniseed, rhubarb, liquorice and peppermint. Because its oil was in demand for making a cordial well known to dram drinkers, as well as in apothecaries' shops, peppermint accounted for more than 100 acres.

Forty years earlier, Lysons[53] claimed, only a few acres of herbs were grown in the parish; but, once the idea caught on, others set up in business, so that an increasing number of people were reaping the benefits of the rich black soil. An article in the *Pharmaceutical Journal* in 1851 commented on the increased area given over to herbs, which then covered more than 800 acres, about 50 of which were lavender. So the physic gardeners prospered on the proceeds of their produce, for which the demand seemed constant.[56]

For much of the nineteenth century, Mitcham was celebrated for its flower farms that perfumed the summer air, "and probably there is not in all the kingdom a single parish on which the wholesale druggists and distillers of the metropolis draw more largely for their supplies", Edward Walford affirmed[54] although in his day the industry was already in decline.

One of the old horticulturalists sang gentle praises of his little garden:

> The jessamine, sweet-briar, woodbine, and rose,
> Are all that the west of my garden bestows;
> And all on the east that I have or desire
> Are the woodbine and jessamine, blush-rose, and briar;
> For variety little could add to the scent,
> And the eye wants no change where the heart is content.[55]

One version of a popular nineteenth century rhyme ran thus:

> Sutton for mutton,
> Carshalton for beef,

Mitcham for lavender,
And Dartford for a thief.

For over a hundred years the gardens flourished, reaching a peak in the mid-century before they began spreading south to Wallington, Carshalton, Sutton and so on.

11
Potter & Moore

Potter & Moore are probably the first names that come to mind in connection with Mitcham lavender. Indeed, they must take a significant place in any account of the trade, since it was their lavender water that made such a resounding hit. It was pronounced as sweet and fragrant when the company celebrated its bicentenary in 1949 as it had been 200 years before.

We know of a Henry Potter, 'a gardener' (fl.1715-1729); and John Potter (d.1742) was 'a physick gardener'. However, everything really began when John's son Ephraim and William Moore, joined forces to erect a distillery, for extracting lavender oil, on the farm on the site of Eveline Road, north of 'The Swan', and overlooking the open triangle of green between the London and Streatham Roads, now known as Figges Marsh. This business passed in succession to James Potter, Ephraim's son, and thence to James Moore, grandson of William and nephew by marriage to James Potter. It continued to flourish in the hands of Moore's son, James Bridger, and only when he died was it purchased by an outside company, W. J. Bush and Sons.

James Potter was a shrewd businessman and an adept nurseryman. So well did he manage the firm that he was considered the country's foremost grower before he died, amassing a considerable fortune in the process. James Moore, or the 'Old Major' as he was known locally, achieved a wide reputation as Deputy Lieutenant of the county and lord of the manor. His claim to this title derived from his purchase of the Lordship of the Mitcham manor of Biggin and Tamworth from the Manship family in 1804. In twentieth century terms he might be considered a small farmer: but in Victorian Mitcham he was a step below the gentry, master in his own village; a source of job, cottage and security, and a colourful and respected figure.

Through constant purchases of land the estate grew rapidly, until, just after the turn of the century, he was cultivating some 500 acres. Considering the value of land so near to the metropolis, amidst

Harvesting lavender in Mitcham.

factories and opulent villas, this must have been a pretty wealthy estate. By mid-century he farmed 543 acres in Mitcham parish, 362 of them which he owned. For some reason, he never seems to have acquired outright possession of the Manor House itself, or its extensive outbuildings.

We get the best picture of Potter and Moore's distillery and farming methods from a survey of Surrey conducted for the Board of Agriculture and published in 1805. Malcolm, its author, was described on the title page of his book as "Land Surveyor to their Royal Highnesses the Prince of Wales, and the Dukes of York and Clarence". He referred to his friend James Moore as "pre-eminently distinguished" in the cultivation of medicinal herbs.[49] Obviously he was familiar with the estate.

In the farm-yard he counted no less than twelve carts of various sizes, five wagons, and a timber carriage. There were sixteen, or four wagon teams, of powerful horses, two or three of which trundled into London every day of the year with herbs, straw or other produce for the market, to return laded with manure and 'night-soil' (household sewage etc.) from St. George's Fields. It was Moore's practice to keep this land in good tilth with twenty large cart-loads to the acre of strong rotten dung from the best London stables.

Ephraim Potter's old house seems to have been demolished some time after the turn of the century, probably by Moore, to make way for his fashionable two-storeyed brick villa. The latter presided over the usual farm buildings, wagon and cart lodges, stables, barns and a counting house or office attached to a dry warehouse. Nearby was a large still-room containing five coal-fired copper stills. Outside the still-room was a large horse-mill, and, a short distance away, a drying-house with coal-fired furnaces which fed hot gases through a system of flues beneath canvas-covered frames. Above them was a loft where the freshly-harvested herbs were stored; when due for drying they were spread on the frames resting on joists about three feet away from the heat.

Moore grew principally peppermint, with smaller acreages of wheat, liquorice, chamomile, damask roses, lavender, spearmint, red roses, pennyroyal, poppies, marshmallow, angelica, hyssop and elecampane, in declining order. Quite a collection, one might have thought, even without the plots of wild cucumber, savin, horehound, wormwood and some arable pasture. Lavender was far from being the largest crop: in fact it seems likely that the lavender boom was not yet under way at the time of Malcolm's survey. Nevertheless, we are given details of its cultivation. In five or six acres of deeply ploughed and amply dressed ground, plants, raised from seed in a bed of fine mould, were planted out at intervals of two feet in the spring, hoed and trimmed in the autumn, then forked round in the winter. Today, we know that lavender makes only frugal demands on the soil, and I am inclined to believe that, if Moore had conducted a few controlled experiments, he might have saved himself the cost of his best rotten dung. It was probably deleterious in such quantities, but, willy-nilly, it was spread on and pointed in every other year.

After a sunny spell in August, when the oil content was found to be at its height, local women gathered the harvest by bunching the flowers, or throwing them into loose bundles of about one hundredweight, called mats. Mats were wrapped in a coarse fibre known as 'bass'. The herb was placed on the mat, flower-head inwards, and the two ends of matting simply skewered together with three bits of wood, resembling clothes-pegs, for carriage to the still-room. Moore paid out one guinea an acre for cutting, and sold the oil to perfumers and chemists at forty shillings a pound.

Moore's unmarried status did not deter him from fathering several children, amongst whom James Bridger, like his father, will always be remembered in Mitcham history. He does not appear to have suffered one bit from the stigma of illegitimacy. On the contrary, his death, lamented in the *Croydon Advertiser* of May 9th, 1885, marked the loss of an exemplary inhabitant, guardian, overseer and churchwarden. He had filled nearly all the parish offices, as well as being a senior director of the Mitcham Gas Company and a large employer of labour in the physic gardens. "As a master, no man or woman, provided they conducted themselves properly, required a second one, and there were at the time of his death, men who had spent the whole of their lives in his service . . . in fact, no man had fewer enemies."

Seven years after the death of his father in 1851, Bridger moved into the Manor House, redolent, as always, of peppermint and lavender. The largest of Mitcham's several distilleries, it was a show-place of the village and a perennial source of fascination to the Epsom racegoers, many of whose coaches stopped at 'The Swan' inn for a change of horses. Here, according to James Drewett, an old Mitcham resident, "could be seen and heard all the accompaniments of village and farm life–the blacksmith, carpenter, wheelwright and repairing shop, the flail and threshing floor, pig yard, barns, store shed, the huge distilling coppers and vats and large drying stoves, the great horse-propelled wooden cogwheel to pump water into the big storage tank which supplied the distillery. From its high position on a barn turret facing London Road, the old Major's clock boomed out hours and quarter hours".[57]

It was the early gardeners who really left their mark on Mitcham history; and W. J. Bush, still growing and distilling herbs in the twentieth century, was, to some extent, supporting a dying industry when he took over the goodwill earned by Potter & Moore. The firm was founded, in Bishopgate, for preparing tinctures and extracts; and Bush, being a man of rounded interests, was quick to realise the importance of the volatile constituents of spices and herbs. Combining his chemical activities in the production of various ethers and esters with his interest in botany and herbs, he pioneered the development of the flavouring-essence industry.

But mergers are part of industrial change, and in 1968 it was announced in the *Mitcham News and Mercury* that the firm of W. J. Bush would suffer the fate of many enterprising young businesses by merging with two other companies to form Bush Boake Allen, the world's largest supplier of flavours and perfumes, with an annual turnover of £20 million. They continued to expand, and soon branches following various lines of research were established all over the world. Even at Mitcham they were extracting nicotine from tobacco as well as distilling herbs. Investigation took place into substances like coconut butter; and chemicals like benzyl alcohol were produced. A far cry, one might think, from the Potter & Moore side of the business, which moved to Leytonstone.

The Mitcham factory was gently phased out, though not without leaving a memory of the firm who first began local operations. In March 1933 the area surrounding the Synthetic Chemical Works was shaken by a frightful explosion; and a record promptly penned in a local Junior School in Bath Road read "in the nearby factory of Potter & Moore". Only the ghosts of Potter or Moore exercised rights over the old factory site: Bush had carried on the work at Bath

Road, yet it was Potter and Moore's names which lived on in the public mind.

In 1968, Albright and Wilson Ltd., of which Bush Boake Allen then formed a part, sold the Potter and Moore section to E. C. De Witt and Co. Ltd., part of the De Witt International Organization, now specialising in the manufacture of toiletries. It is a source of pride to the parent company, to which their headed writing paper bears witness, that they incorporate a firm established in 1749. As recently as November 25th, 1970, if *The Guardian* is to be believed, it was not too chilly for two lasses, wearing faithful copies of the original sellers' hats and dresses, to re-enact the old street-selling custom outside the home of a famous customer, Samuel Johnson, in celebration of the 222nd anniversary of Potter & Moore.

12
The Industry Spreads

As the profitability of herb growing increased, so its popularity grew. An entry appearing in the Mitcham Vestry Minutes for 28th March, 1802, seems to be an appeal by nine parishioners against full payment of tithes because of an earlier agreement to pay the clergyman twelve shillings annual rent for occupying a garden ground. Over the past twelve years there had been "a very great increase in growing ground therein which before was used as meadow pasture or arable", and thirteen more parishioners had turned herb grower.

Trade directories give a useful if erratic idea of current employment trends. Pigot's[58] for 1825 mentions several hundred acres of physical gardens, though no actual gardens are listed until 1832, when there are five; whilst Robson's *Commerce of Surrey*[59] gives thirteen in 1838. Since Pigot's only lists six names in 1839, though Robson's sticks to thirteen, one supposes the first was more selective in its choice. Post Office Directories, too, tend to estimate lower. From 1845-66 their number varies, though it is never more than seven. From 1878 onwards, however, Kelly's and the Post Office Directories[60] mention only Mr. Bridger, besides a number of market gardeners who must presumably have used part of their land for herbs. In 1903, the Post Office Directory notes that cultivation of lavender and roses had to a great extent ceased; and thirty-five years later all mention is omitted from the Kelly's Directory.

At the height of herb growing the *Pharmaceutical Journal*[61] mentions six growers from Mitcham and its environs: Moore, Arthur, Martin, Newman, Sprules and Weston. Later on, Holland Jakson and Miller were other principal growers. The first five shared fourteen stills scattered throughout Mitcham. One was at the corner of Killicks Lane; one at Commonside West, next to Cold Blows; one each at Sherwood and New Barns Farms; and one where 'The Beehive' now stands, according to Mr. J. D. Drewett.[57] But when he volunteered the information in 1923, they had long ceased to exist. At one time they were all emitting the coarse smell of unrefined oil as they poured black smoke into the Surrey sky.

We know that Mr. James Arthur, son of Richard Arthur, also a physic gardener, had a farm to the south-east of Mitcham Common which employed a large number of men and women, growing solely herbs. Long after it had ceased being used by the family, the distillery, then belonging to the French firm called Jakson, gathered grime in the Croydon Road. Slater was another family so involved in the industry that when the last of the Mitcham growers, a Mr. Steward Slater, died in 1949 aged 91, his family had been in the business for over 200 years, living for the last century in a wood-framed cottage in Love Lane. One of the Slater brothers, William, emigrated to Australia, where he built a still for extracting lavender and eucalyptus oil; another, Benjamin, wrote a booklet about the Mitcham herb gardens.

If it was largely as distillers of peppermint that Messrs. J. & G. Miller gained their reputation. There are some early twentieth century photographs to prove that they grew a variety of herbs, including lavender, in their fields on what is now the St. Helier Estate. The distillery at Beddington Corner was situated north of Wood Street and parallel with York Street. It was taken over by Hollands Distillery when George Miller died.

The first photograph shows a cart stacked high with sausage-shaped mats or bundles drawn-up alongside their distillery. Two men in dark waistcoats and baggy trousers, caps and rolled shirt sleeves, stand aloft, preparing to tumble the harvest through a doorway on the upper floor. The next is of the distillery itself, an unprepossessing two-storeyed corrugated-iron shack, with a tall, blackened, brick chimney towering above heaps of spent herbs, while another catches the scene inside the top storey where two men are stamping down the herbs in a circular copper still, projecting from the ground floor. Two others similarly clad, and with handsome moustaches, are untying bundles to add to the increasing load, under the direction of a supervisor complete with beard and watch chain: surely none other than Mr. George Miller, senior partner and, we are told, an ardent supporter of Joseph Chamberlain's Tariff Reform Committee.

This brings me to a fourth photograph, of a worker pouring oil from a can into a measured metal Winchester quart, under the eyes of both senior partners; and the last, showing a long room in the

The distillery at Mitcham, owned by Messrs. J. & G. Miller ABOVE LEFT: Unloading bales from a cart. ABOVE RIGHT: Exterior of the distillery. BELOW: Treading down the herbs.

ABOVE: Pouring oil from a can into a measured Winchester quart (Millers' distillery).
BELOW: The senior partners sampling the quality of the oil.

distillery furnished from end to end with a wooden table. Both walls and table are lined with uniform bottles of oil, and this time the company directors themselves are sampling its quality.[62]

The partners of the firm were two brothers whose father had founded it forty years earlier. The father had been chiefly a grower, while his sons established the distillery; but their claim to be the proprietors of the oldest and largest stills in the country is justified by their having acquired some of Potter & Moore's original equipment. Millers actually controlled 1,200 acres of herbs, supplemented by market gardening and fruit culture, paying out nearly £10,000 a year in wages, which made the enterprise more than double the size of Potter & Moore's at the time of Malcolm's review.

The brothers' four large old-fashioned pot stills ran day and night without break, apart from Sundays. Herbs were covered with water to within two or three feet of the top of the still, and after about half an hour's heating, when the distillate began to come over, the fires would be damped down. At the end of six hours, distilling was complete, and the steaming herb would be cleared – a sweaty job we are told – for which the men stripped to the waist. Eleven or twelve pounds of oil was a fair average yield from a distillation of about sixteen hundredweight of dried herbs, but much depended upon their condition. Hollands Distillery continued to trade under the old name. Their speciality was what they described as 'terpeneless oils': bays, bergamot, cloves, geranium, lavender, lemon and so on. They still, in fact, distil at 37-9 Wood Street in more modern buildings, but no longer Mitcham herbs.

13
Wallington, Carshalton and Sutton

Sprules is an unusual name, and those in the London area seem to be related; they may well have originated in Wiltshire. Our interest in the family begins with William, born at Mitcham in 1811. By the time he was thirty, we know that he had a wife and seven children, six of them girls, to bring up on the proceeds of a herb farm at Reynolds Mill in Carshalton. The mill was next to the present Reynolds Close, about a quarter of a mile west of Hackbridge Station, on an island in the River Wandle. It was called, variously, Hackbridge Mill, Culvers Mill, or Carshalton Mill.

A decade later, William was employing twenty labourers on twenty-four acres at a farm in North Cheam, with fields of lavender and peppermint at Cheam, North Cheam, Sutton Common and Beddington Lane. By this time there were two more children to add to the increasing brood, and William kept expanding the business. Another ten years found him with eighty-eight acres in the Mitcham region (a conservative estimate: he may have surrendered some of his land after the Carew estates were broken up). His farm, known as Lower Homestead, was rented from the Carews. It lay immediately south of Beddington Lane Halt on land now covered by sewage works. From lavender to sewage in a hundred years is the price we have paid for progress!

William's final move, some time about 1864, was to Melbourne Road, Wallington, where his lavender stretched from Boundary Road to Manor Road, and south of Wallington Station.

The eldest son seems to have left home to make his own way in the world, but three daughters followed in their father's footsteps, growing and distilling herbs. Sarah took over the business when her father died at the fine old age of eighty-two, and, through her energetic endeavours, managed to revive the industry after a series of bad seasons had seriously reduced the crop. Melbourne Road was then isolated in the midst of agricultural land; although, not long after, it became surrounded by Victorian houses, except for a market garden plot immediately adjoining the Sprules' premises. Number

Miss Sprules in the garden of 40 Melbourne Road, Wallington.

40 stood back from the general line of houses on the south side of the road, backing on to the West Croydon-Sutton railway line. Those were days when everyone knew almost everyone; and Miss Sprules, and her old distillery which was kept active until her death in 1912, are still remembered by some inhabitants of the district.

Mr. K. A. Pryer of Carshalton remembers Melbourne Road when: "There was thick housing on the far, southern side of the railway for at least half a mile before one reached farmland, and that did not then grow herbs".[63]

Miss Sprules herself is remembered as a charming, very busy old lady, who once walked through her fields with Queen Victoria. She was, by special appointment, 'Purveyor of Lavender Essence to the Queen';[64] and her lavender water won medals at exhibitions in Jamaica and Chicago. She also grew roses, chamomile and two sorts of peppermint: the black, which has purple stems, and the white, with green stems, which yields an oil of greater delicacy. Her Wallington distillery gave employment to a number of thankful villagers. Apparently, Miss Sprules allowed her labourers to take away the surplus peppermint water which ran out of the stills, as a token of gratitude for their endeavours. They bottled it to use for medicinal purposes. She also provided pocket money for several poor women who were employed in selling her dainty wares: fragrant essence, lavender salts, sachets and faggots, on commission. Miss Sprule's endeavours, it was recorded in the *Girl's Own Paper* of 1891, were more profitable than the general run of women's employment.[65] I regret that was not saying very much.

There were other similar enterprises in the district. Mr. Edward Martin of Ewell rented and worked a farm at Nonsuch, Cheam, on land which is now part of the public park. It was from this farm that plants were obtained to start an industry at Westcott, near Dorking.

John Jakson, too, grew chamomile, mint and lavender at Little Woodcote Farm, Wallington, until 1914. An Inventory of Valuation[66] for this farm, dated January 1912, gives the following acreages out of a total of 250¼ for the whole farm: "White Mint 12¼, Black Mint 40¾, Ground prepared for mint 29, Chamomile ½, Lavender 2¾."

A writer in the *Daily Mail* of August, 1914, recorded his search for Surrey lavender, with a tramp along two or three miles of winding country road, the "hedgerows gay with poppies, campions, and fool's parsley, to find it". And there may still be local residents who can remember the sight of field upon field of blue, contrasting with huge white sheets of camomile, extending from Carshalton up to Wallington station which was then a small structure covered with roses.

Mr. J. Smith, then of Stanley Road, Wallington, told me in 1972 that "it was all lavender" down by Queen Mary's Hospital as well as at Barrow Hedges Farm, where the Primary School stands. He knew Mr. Woods, the owner, whose farm stretched right up to Oaks Park, from where he sent waggon-loads of the flowers to be distilled each day at Leatherhead; although there was a distillery for mint and lavender at Shorts Farm in Shorts Road, Carshalton, where the Express Dairy now operates. Mr. Woods had farmed in Carshalton for 60 years; he owned 20 acres of some of the finest lavender in England, and he learned the secrets of successful growing from his father before him.

Mrs. Hilda Machell and her husband lived at Drift Bridge Farm, and worked land near Oaks Park that was bought up in about 1920 for the smallholdings. Peppermint was the mainstay of their farming, and one of Mrs. Machell's earliest memories was, as a child of about three, clambering up a bank surrounding a field of growing peppermint in Carshalton, and falling flat on her face in the waving fronds.

Mr. and Mrs. Machell remained at Drift Bridge until about 1965. Many years of their lives were closely involved with the cultivation and distillation of herbs. Comparing their twentieth century farming methods with those of Potter & Moore in 1805, Mrs. Machell wrote: "I find it so interesting that the methods have changed so little. How much easier it must have been regarding labour however!

We did not get our well rotted dung from St. George's Fields, but we had horses in the early days and always kept a goodly sized herd of pigs and latterly, cattle. We also grew lavender. (A wonderful sight indeed were the fields of great bushes of lavender.)" [67]

Harvesting lavender in Carshalton; the trees in Oaks Park can be seen in the background.

14
Cultivation

One stern critic of Mitcham growing was Mr. J. Ch. Sawer (as he signed himself), a pharmaceutical chemist who also grew and distilled on a small scale near Brighton. Because lavender cultivation was new to Sussex, and he was notably successful, his work attracted attention in pharmaceutical circles. He took his lavender growing seriously, not only experimenting with different materials and techniques until yields and quality were as high as he could achieve, but studying the market until he had a thoroughly comprehensive view of the scene. His conclusions were dismal: lavender was in the doldrums. Since the late 1880s, when disease in the Mitcham fields caused the price of English oil to escalate madly, the market had become inundated with inferior French oils. The public would buy anything cheap, he lamented, however bad, and the majority of chemists and perfumers continued to use foreign oils even when the English price fell. Only the United States consistently supported the British market.

Instead of the true non-camphoraceous *L. officinalis,* the Mitcham growers seem to have cultivated hybrids which contained camphor. The result was that much of the Mitcham oil was excluded from the highest grades of perfumery. At the same time, much of the oil prepared in this country was superior to that prepared in France. Certainly during the nineteenth century it commanded a much higher price: "as many shillings an ounce as the French oil is worth per pound".

When lavender was still propagated by seed, it was sown in March, and the plants were transplanted, when they were about two inches high, to nursery beds, where they remained until the following year, when they would be planted out for good. By mid-century however, it seems to have become more usual to propagate vegetatively, either by root division into three or four new plants, or by cuttings. Since hybrid plants do not breed true, more uniformity would have resulted from these methods. Initial planting was slightly closer than Potter & Moore's at the time of Malcolm's survey: eighteen inches apart, in rows nine inches apart. A catch crop of lettuce or parsley was sometimes taken; but, as the lavender matured, alternate rows were removed, until the final stand was three feet either way, without any intercrop.

Sawer recommended longer productivity with much wider planting: four feet apart in rows five or six feet wide. One reason for this was to increase the plants' flower-bearing capacity, another was to prevent disease. Although dieback was rarely, if ever, seen in gardens where single plants were grown, it occurred from time to time in the middle of plantations. Sometimes in the Mitcham area its effect was devastating.[68] At first, it was believed to be caused by some sort of 'poisonous influence of the flowers' excessive aroma', then by frost or lime deficiency. But W. B. Brierley's investigation in 1916[69] provided proof that it was undoubtedly *Phoma lavendulae,* or shab, a parasitic fungus known to have dogged lavender growers periodically ever since lavender was commercially cultivated. Further work on the fungus, carried out at Cambridge, elucidated some points about this hitherto baffling disease.

The first definite symptoms of infection appear at the end of May, or the beginning of June, when young shoots in isolated patches turn yellow, wilt and die. The disease extends downwards; killing the plant, and, at an advanced stage, tiny black heads of spore-producing fruiting bodies are revealed, hidden amongst the hairs of the stems. It may spread to larger areas; in fact a bad epidemic will sweep through entire fields, encouraged, Sawer suggested, by careless scraping of the main root with a hoe, or by excessive manuring. His first suggestion was correct. The fungus invades a healthy plant either at the leaf axil or through any fresh wounds, such as those made when the crop of flowers is cut, or when plants are trimmed. The disease was commonly spread by propagation with infected cuttings.

There were other hazards entailed in growing lavender; we have abundant evidence of this. One year a dry spring might delay planting young stocks; older crops would be thinned by drought; and the wells would run dry, leaving the stills short of water. Although scarce, the oils would be fine and volatile, so the price would rise and growers benefit; but the shortage of crops would lead to competition for work, and labourers would not fare so well. The following season might be equally unpropitious for quite different reasons, such as excessive frost and cold.

Agriculturalists are proverbial grumblers: their forecasts are expected to err on the pessimistic side; and one always feels that there was apprehension over prospects which could easily deteriorate. Crops might be ravaged after wind, rain and cold; or hitherto unbroken expanses of colour might become so blighted with disease that there would be almost no harvest at all. There were years when lavender was abundant and the quality good; but circumstances were seldom right for lavender's rather specific requirements.

Relatively low rainfall is of paramount importance. Dry weather during June hardens growth and helps to put the plant's energy into bloom instead of foliage, whilst later rains encourage the flower heads to rot and drop, instead of maturing properly. Heavy and prolonged rainfall any time during June, when the spikes first appear, until the crop is harvested in July and August, can result in a loss of as much as 50% of the crop. It is only necessary to look at the soil beneath the plants, heavily sprinkled with fallen flower heads, to know that the harvest will be a poor one.

All these reasons contributed to the contraction of the lavender and herb industry in this country. "It almost appears as if the herb cultivation in Mitcham were doomed to extinction within a measurable span of years", a pharmaceutical journal predicted unhappily in 1891.[70] "With all its advantages it is questionable whether the Mitcham district is not too near to the metropolitan area to remain a suitable ground". This foreboding was followed by a hopeful plea that lavender should be cultivated in other parts of the country, since it was too much a part of the popular 'English' image to be allowed to disappear completely.

Critics put the rank appearance of lavender plants down to impoverished soil resulting from poor management. It was more than this. It was lack of scientific principle, and unorthodox, somewhat hit or miss, methods of cultivation and distilling that had not been improved upon for at least fifty years. Apparently the distillate was separated into two grades, working on practical experience rather than chemical calculations, and the results were anything but uniform. Small wonder that much genuine English lavender oil failed to comply with pharmaceutical requirements.

15
From Herb to 'Essence'

The harvest was cut by men and women using small bagging-hooks made in Mitcham. The erratic school attendance of girls employed in the fields on rosebudding, fruit-picking or gathering chamomile during the season, suggests that children may also have helped with the lavender harvest. Annual reference to absences occurred between 1863 and 1877 and, by 1889, the authorities in one Mitcham school had become so worried by consistent absences that they offered prizes for regular attendance. Even in this century, schools in the lavender areas shrank during harvest time, when their pupils were either working, or taking lunch to relatives in the fields.[71]

But adult labour, too, was needed in and around Mitcham at harvest time. Because the lavender and peppermint crops generally coincided, W. Newman Flower reckoned (in *Country Life,* 13th July 1901) that several hundred extra hands were required. Apparently, whole Irish families, from parents to youngest children, came over to Surrey and subsequently made their way into Kent in time for the hop-picking.

It was the practice to lay the flower-heads inwards on the mats when harvesting, to protect them from the sun. When the bundles arrived in the still-room, the three wooden skewers which secured them were simply pulled out and chucked on a heap in a corner, before the contents were thrown, stalk and all, into the still. Sometimes, however, women were employed to strip the flowers: incredibly laborious work that took twenty or thirty of them six or seven hours' hard work to make up the half-ton needed to fill a still.

The principal growers distilled their own produce as well as that of smaller proprietors. Others grew very little themselves, dealing almost exclusively with the produce of small growers, to whom they charged a fee of about twenty shillings per load for a medium-sized still. The principle behind distillation is basically so simple that it allowed little variation; but records gathered over the second half of the nineteenth century show that performance, distilling times, size of load and other factors differed considerably. However, these statistics may have varied in accordance with the accuracy of the reporters and the honesty of the growers.

The distillery was usually a two-storeyed wooden building with a tiled roof, and access to the upper floor by steep wooden ladders. The stills themselves, resting in solid brick furnaces which reached to the upper floor, were huge perforated copper containers fitted with a massive copper head. Sometimes schoolboys were employed to tread the flowers in, and sometimes they got stung by bees in the process; but after a few days on the job, we are assured, they became practically immune to the pain. It is on record that workers were once paid 'danger money' for each sting received – but precisely when or where is not stated.[72]

The still was filled with boiling water, and its head sealed on with clay, before the fires were lit so that oil-laden vapours could pass into a coiled 'worm' 200-300 feet long, its end immersed in a wooden vat about ten feet high and forty to fifty feet in circumference. There the vapour condensed into a copper container, oil and water separating. The Mitcham stills were the largest in the country, and very much larger than those used in the flower farms of southern France.

Their capacity seems to have varied between 700 and 1000 gallons. They took a couple of hours to get up steam; and the finest oil was drawn in the next two to two and a half hours. A total distillation of two or three distillates took about six hours. An 1874 survey gives a yield of one pound of oil from seventy pounds of flowers, an optimistic and, I would say, unreliable figure, if compared with Miller's eleven to twelve pounds from sixteen hundredweight fifty years later–which turns out to be just about half as much. Other general estimates of a pound of oil from 112 pounds of flowers in a good year, and half this in a poor one, and the present Norfolk figure of roughly 225 pounds of herbs to a pound of oil, imply that the early distillers squeezed the very dregs from their material.

The cooked residue was cleared by two men working in unison. One handled the crank of a windlass from which a chain ran, passing through a hook in the ceiling and fastening to the end of a heavy six-pronged fork manipulated by his partner. At a signal from the latter, down went chain and fork to secure the smelly residue; the chain tightened, and a forkful would be brought to the surface and thrown through a trap door into the yard below. It was used for mulching and manuring; although it is without any value, so it can have done

Lavender still at Mitcham from Piesse 'Art of Perfumery' (see Ref. 46, p. 107).

little good, beyond suppressing weeds. The same stills, of course, were used for other distillations, and, where lavender followed mint or chamomile, it was necessary to boil out the still with lime and water, as well as washing the 'worm'. The oil was bottled into four-or five-pint Winchester quarts, where it was kept uncorked to mature. Spent bushes were ploughed up to thatch the sides and ends of sheds, or stored with furze from the common to fuel a spanking Guy Fawkes blaze. The Rev. Henry Ellacombe has told in his *Plant Lore and Garden Craft* (2nd ed. 1884), how the air would be fragrant for miles around when Mitcham growers were burning the old bushes.

The drying house, with lavender suspended from the roof, in a Mitcham distillery. From Piesse 'Art of Perfumery' (see Ref. 46, p.107).

16
On the Wane

At the beginning of the nineteenth century the population of Mitcham was 3,466. By mid-century it had grown by just over one thousand and, at the end, it had quadrupled. Within a hundred years, onto the quiet rural village straggling along the London Road had been grafted a modern suburb many times its former size, carrying with it the advantages and disadvantages of metropolitan life. In the 1920s and 30s, residential properties mushroomed: in the words of Arthur Mee, London had "crept haphazard to its gates . . ."[73] Then, in 1934, Mitcham was granted a Charter of Incorporation as a Municipal Borough.

As the village stepped out of its thousand-year-old country bumpkin ways into svelte urbanisation, lavender naturally fled. The trend had started well before the turn of the century. Thirty years earlier, Wallington was called the headquarters of lavender growing, and blue fields sprouting in Wallington, Carshalton, Beddington, Waddon, and Sutton made the train journey from West Croydon to Sutton a colourful experience. Fields of lavender could also be seen in the Crystal Palace area. But everywhere the story was the same: no match for the speculative builder, the smallholder was being driven out.

Urbanisation alone was not responsible for the break-up of the old herb gardens. Many of the physic gardeners turned over to market gardening; and grew chrysanthemums and vegetables, which flourished for years, when glass allowed scope for a more intensive industry. As we have already mentioned, by and large, the Mitcham lavender industry was unsophisticated: moreover there was competition from overseas. Compared with forty shillings a pound for oil, given in Malcolm's agricultural survey of 1805,[50] prices at the end of the century were sixty shillings, eighty shillings, ninety shillings and, sometimes, a hundred shillings, rising in 1881 – a black year – to as much as two hundred shillings. Meanwhile French, Dutch and Japanese oils were being imported in vast quantities, to retail at around eighteen shillings a pound. Vendors of French oil started to adulterate the English oil with it, until, by degrees, the latter became almost entirely replaced.

Another reason for the decline of the physic gardens was the advances made in synthetic organic chemistry which made possible the production of man-made drugs.

"Those who are engaged in the harvesting of lavender have verily a pleasant occupation",[74] enthused a *Chemist and Druggist* reporter sent down to comment on the Mitcham industry: "Why this crop is not more extensively cultivated we are at a loss to determine". But he saw only one side of the picture, and that was a rosy one. Set against the pleasures of open-air indulgence were the vicissitudes of a life largely dependent on the weather, susceptible to extensive loss from marauding, and dogged, from the labourers' point of view, by prevailing social and economic conditions. Romantic though the fields and blossoms appear to have been, there was much hardship amongst the workers.

Farm labourers toiled for about fourteen hours a day, at a wage of no more than ten to fourteen shillings if they lived near London; further south it was as low as seven or eight. Tom Francis, a Mitcham inhabitant whose collection of manuscripts, photographs and lantern slides is preserved in the local library, gave the rate in the area as ten shillings and sixpence to twelve shillings and sixpence a week.[75] In the early part of the century, he said, thirteen and a half per cent of the entire population of Surrey consisted of paupers, to whom employers paid as little as possible, leaving the Poor Rate to make up a subsistence wage.

Fortunately, because the men working on farms and in the distillery fields were unable to do anything at all in the winter if the land was frostbound, landlords would only accept rent from them during the summer. In fact, one redeeming feature, before the advent of the Welfare State, seems to have been the charity and good nature of some of the wealthier, land-owning members of the community. In Mitcham they provided relief by running a soup kitchen twice a week. Those with a subscriber's ticket would get four ounces of real meat and vegetable soup for one penny, and children got away with as much as they could stuff themselves with on the premises. Boys and girls were often obliged to pad barefoot to school, yet the odd donated pair of boots might well be promptly pawned.

One bunch of flowers that did not reach the distillery, in August 1886, was, according to the *Wallington Herald*, presented with pleasure to the little Princess Louise. Many more did not reach the distillers because they were pilfered. Thieving was often reported in the local press. Complaints from the growers, during August, of wholesale thefts from their fields were sometimes met by protection from police officers over the Bank Holiday, after which youths were hauled before the Croydon County Bench and given petty fines.

. . . . . .

Sutton for mutton,
Carshalton for beef,
Croydon for a pretty girl
And Mitcham for a thief!

What has happened to the old rhyme? Mitcham doubtless owes the less flattering version to gipsies and vagabonds who frequented the Common, earning it a bad reputation. In fact, whenever possible, travellers avoided the Surrey commons, all notorious for lawlessness. But lavender has been dropped from the rhyme. Requests for the small curiously-curved sickles, or mint hooks, used for its harvesting, were still being received by local ironmongers, but the plant itself had almost disappeared.

One of the last growers, Mr. Henry Fowler of Bond Road Nursery, became something of a celebrity, being constantly interviewed by the press in the first quarter of this century. A portrait of him in the *Daily Chronicle*[76] shows a fine, weathered face, capped and bearded, above great bundles of the lavender he sold at Covent Garden. In 1910 he reckoned he was the only grower/trader left in Mitcham. Ten years later he thought he was, perhaps, too old for the job; but it seems he was supplying occasional orders for another five years, until he died. This left William Mitchell, another handsome old-timer of 94, as the oldest survivor. At his death three years later, few remained to reminisce, from inside experience, about the lavender and herb industry.

As, one by one, they aged and died, the passing of each old grower sparked off a flurry of nostalgia. Henry Fowler, William Mitchell, the Slater brothers and Augustus Stanley, all born in the heyday of lavender: men who had spent their school holidays in the chamomile fields, some later becoming well-known figures behind their Covent Garden stalls. Meanwhile, lavender travelled south. The last fields

William Mitchell 1831-1928.

Henry Fowler 1846-1925.

within the boundary of Mitcham to be disposed of, those owned by Mizens in Eastfields, went in 1933; and in Carshalton the last fields (owned by Bush and Company) were sold to the London County Council for housing in the 1920s.[77] The plants, however, were transferred to Epsom; Pole Hill, Sevenoaks; and Wallington, so that Mitcham lavender, flourishing outside the borough, furnished material for the company's Mitcham stills. An old still-book spanning twenty-two years, from 1911 to 1933, shows the sources for the distilling materials becoming increasingly further afield.

Lavender seller at Temple Bar. This is the picture described on the opposite page as the Phillips plate.

17
Criers and Their Cries

Since John Lydgate, a Benedictine monk of the 15th century, first acquainted us with the little rhymes sung by milkmen, bakers, chimney sweepers and other itinerant traders to attract the attention of customers, the cries of London have been a stock subject for nostalgic reminiscence. They were borrowed by the early dramatists; adapted for children; set to music by eminent composers – Thomas Weelkes, Orlando Gibbons and Richard Deering; and reproduced in illustration by artists and printers, in various forms from the rudest woodcut blocks to the finest of copper and steel-plate engravings.

The early cries were a colourful and motley collection, courting custom for almost any commodity from singing birds to three rows of penny pins; a dish of eels; a fork or a fire shovel; fresh oysters, of course, and fair cherries – even 'fresh cabbage' – but not, as yet, lavender.

Our first documented evidence of the lavender cry seems to be in the British Museum's Roxburghe collection of *Ancient Songs and Ballads,* printed around 1700. To the tune of *The Merry Christ Church Bells,*[78] *The Cries of London* incorporates an exhaustive list of London ditties in twelve-line stanzas, each ending with the jaunty refrain:

> Let none despise the many, many cries,
> Of famous London town.

All the old favourites feature in the song: rosemary, sage, thyme, fine green mint and lavender; although the last has only one slightly mundane line: "Here's fine lavender for your cloaths".

The Phillips plate illustrating a supplement to *Modern London* (1805)[79] is the earliest print of a lavender seller. It was later popularised by Player's cigarette pictures, as No. 3 in the first series of twenty-five. Against a background of Temple Bar, a lavender seller is depicted in traditional costume. An explanatory caption on the lavender plate reads:

" 'Sixteen bunches a penny, sweet lavender' is the cry that invites in the streets the purchases of this cheap and elegant perfume. The distillers of lavender are supplied wholesale and a considerable quantity is sold in the streets to the middling classes of

The original painting of a primrose seller by Francis Wheatley R.A.

Yardley's adaptation of Wheatley's original.

inhabitants who are fond of placing lavender among the linen yet unwilling to pay for the increased pungency of distillation."

One of the fourteen paintings representing the cries of London by Francis Wheatley has been immortalised by Yardley of London, who adopted it as their trade mark for all lavender products in 1913. Porcelain figures were also modelled of a lady with two children bearing baskets of lavender spikes for sale, but the Yardley version is to some extent fudged, since primroses, not lavender, were drawn in the original.

The cry was not written down until Lucy E. Broadwood and J. A. F. Maitland edited *English Country Songs*, published in 1893:

> Will you buy my sweet lavender,
> Sweet blooming lavender,
> Oh buy my pretty lavender,
> Sixteen bunches a penny.

By this time there were three shorter versions of the words and many variants of the tune. One variant was embodied in classical music when Vaughan Williams wrote his London Symphony. The modern lavender call with its falling chimes of melody is sufficiently like the Rosemary and Bay cry in Deering's *Street Cries Madrigal* for folk song enthusiasts to suspect it is descended from an earlier one for rosemary. It may in turn, they suggest, have inspired the refrain for an old revival hymn called 'Happy Day': the words 'Happy day' taking the place of la-ven-der or will-you-buy.

Who were the lavender sellers? Apart from the market-gardeners or small-holders themselves, hawking Mitcham lavender was a gipsy prerogative. Mitcham's association with gipsies goes way back to the early days of its August fair, an annual festival important in gipsydom, supported by a colony who congregated, and to some extent settled, in the area. There were grades of hawking like everything else. Because lavender selling did not rate very high, and gipsies were amongst the least affluent, it was left to them.

Part of the gipsy's job was to make up bags of lavender for chests and wardrobes; smaller ones for young middle-class ladies to wear between their breasts; and flat, detachable sachets for placing under the armpits, where they replaced the effective but uncomfortable pieces of india-rubber, sewn into the arm seams by dressmakers to prevent perspiration being absorbed by clothing.

Watercress from the Mitcham Wandle beds was another familiar gipsy ware. It was sold to small greengrocers to retail to costers, who in turn hawked it in the poorer London streets. At the same time, of course, it was also sold by the growers direct from the beds. But, at about the time lavender finally faded out, an epidemic of typhoid in the Croydon area, believed to have been caused by polluted streams, halted the sale of watercress.[80] I have one version of the street cry coupling lavender and watercress:

When Pol stays here and Jack goes there
To earn their provinder
Her cries are all in Bethnal Green
Sweet lavender, sweet lavender
Who'll buy sweet lavender?
And oft she wonders if her Jack
Enjoys a man's success
Who cries on top of Stamford Hill
Young watercress, young watercress
Who'll buy young watercress?[81]

Street sellers were not unreservedly welcomed. Forbidden under a London by-law to ring the bells or knock at the doors of houses, every one of these wandering characters depended solely on his or her withering cry, a cry that must rise above the rattle of iron wheels on cobblestones, drown other criers, and penetrate every garret and back room. One of the troubles was that the town criers were just too noisy, and residents protested against vendors vying one with another in a loud and repetitive din. "Pleasantly suggestive as it is to some of us . . . their cry would probably drive away a flea with a musical ear", was a caustic comment from an early twentieth century *Mitcham News*.

Yet, if the lavender cry was noisy, it remained one of the most popular and enduring. It was certainly sung long after sixteen bunches to a penny became an uneconomic rate for the seller. William Stewart explains in *Characters of Bygone London* how the issue was avoided by a discreet slurring of the appropriate line. He gives a version that was calculated to persuade:

Will you buy my sweet lavender, lady?
Only 16 bunches for a penny,
You'll buy it once, you'll buy it twice, lady
It'll make your clothes smell, oh, so very nice.

A well-known gipsy at the Mitcham Fair with the name of Sparrowhawk made no bones about inflation:

Who'll buy my Mitcham lavender...
It makes your handkerchief so nice.
Who'll buy my blooming lavender...
Sixteen bunches for a shilling.

Dresden flower seller group, a ceramic interpretation of Wheatley's painting.

The late Mrs. Rebecca Jeffries and her daughter, Mrs. Rachel Butcher of Love Lane, Mitcham, had a store of old cries from their lavender-selling days. Mrs. Butcher said in 1954 that she started selling on her own when she was sixteen, but none of her daughters planned to carry on the job.[82]

Mr. Stewart tells us that the sellers tended to be young women with sleeping infants bound to their side, leaving their right arm and hand entirely free. If a babe was not available, a small bundle of clothes might be used as a substitute; for gipsies, being shrewd judges of human nature, knew how to arouse compassion. Using a small ragged boy was another ruse that seldom failed to attract a few sympathetic coppers, of which the urchin would no doubt be promptly relieved, as soon as his benefactor turned the corner.[80] I am inclined to believe that Dorothy Davis's mournful portrait of the early criers is a fair one: " 'Won't you buy my sweet lavender' sounds plaintively romantic when sung in a drawing room and looks pretty enough when depicted in a modern cosmetic advertisement. But there can have been little romance or prettiness about the original street-criers of London."[83]

Even contemporary children's literature urged boys and girls to spare a coin for their less fortunate brethren.

On August 13th, 1910, *The Chronicle* attributed a "horrid nasal cry" to Mr. Fowler, the veteran Mitcham grower, who was still claiming to dispose of large quantities of his crop at Covent Garden in the season, though there was little growing in Mitcham at the time.

Four years later the loss was lamented of a little old lady in print frock and spotless apron, who used regularly to appear in the Strand with a large basket of bottles of lavender water of her own distilling – strong and oily at one shilling a bottle.[84] Throughout the 1920s and into the 1930s, autumn was heralded by at least one record in the Mitcham papers of the old cry. On one occasion its source was identified as a woman dolled up in a shoddy blue costume with orange sweater pulled skin-tight across an ample bosom. She was retailing bunches for half a crown.[85]

Soon after this Mrs. Sparrowhawk sang her refrain for the BBC; and a London seller of fifty years standing, known as the 'Lavender Prince of Westminster' gave his cry into a microphone after a fortifying session of beer.[86] Right into the mid-1950s one resilient resident from Mitcham's Love Lane was plying her trade with her basket in Notting Hill. One very late bird was spotted in St. James' Church Garden, Piccadilly, with an armful of lavender and was recorded in 'Mitcham Notes' in the *Mitcham News*[87] in 1958.

. . . .

Come buy my lavender, sweet maids
You cannot think it dear;
There must be profit in all trades,
Mine comes but once a year.

Just put one bundle to your nose,
What rose can this excel;
Throw it among your finest clothes,
And grateful they will smell.

Though Winter come, it still retains,
The fragrance of today;
And while the smallest part remains,
Your pocket will repay.

One penny's worth is all I have,
This sold, my stock is done;
My weary footsteps you might save
By purchasing this one.

This was culled from a children's book called *London Melodies* (1812). It does not have quite the brightness, the spanking vitality, of the actual cries. But approaching nearer to them in flavour is a song from Cumberland Clark's *Flower Song Book* (1929):

Lavender, sweet lavender;
Come and buy my lavender,
Hide it in your trousseau, lady fair.
Let its lovely fragrance flow
Over you from head to toe,
Lightening on your eyes, your cheek, your hair.

Probably the commonest version of the real cry, as opposed to poems and songs, is simply:

Sixteen good bunches a penny! Blooming lavender.
Blooming lavender!
Who'll buy sixteen good bunches a penny?
Blooming lavender! Lavender!

This one is a close runner up:

Here's your sweet lavender, sixteen
Sprigs a penny,
Which you'll find my ladies, will
Smell as sweet as any.

Another, which specified Mitcham produce, was:

Ladies, come, make no delay
My lavender fresh cut from Mitcham, and I am round here today.

Ladies buy a pennorth of my lavender,
There are 16 good branches a penny, all in bloom.
Some are large and some are small;
Take 'en in, show 'en all,
There's 16 good branches a penny, all in bloom.[88]

But folk song enthusiasts, taking down versions from genuine hawkers, found there were almost as many tunes and words as there were lavender sellers.

Lavender sellers in the 1860s, from a printed version in the Guildhall library.

18
Further Afield

In 1891, the principal lavender plantations were reputed to be at Mitcham, Carshalton and Beddington in Surrey; Hitchin in Hertfordshire; and Grove Ferry near Canterbury in Kent. At Patcham in Sussex, lavender growing was already on the wane.[68] By the end of the decade, Hitchin was the most important centre, whilst at the other end of the scale, cottagers in Worcestershire were ekeing out their earnings by selling lavender water made from the produce of their gardens. They gathered the flowers and sent them mixed with rosemary, roses, musk and thyme, to the nearest town to be converted into 'sovereign water'.

In Derbyshire cultivation had ceased by the end of the nineteenth century. Then, in 1910, Dorset suddenly appears on the scene. An article in *The Times* gave Mitcham, Hitchin and Dorset as the principal areas of production.

Doubtless there were many other parts of the country where the crop might have been successfully grown, but marketing would have been a problem since crops deteriorated in transit. Distillers were loth to accept lavender from any distance, which meant it was difficult to dispose of, unless a grower possessed his own still. On the other hand, most large towns provided a fair market for bunched flowers.

Canterbury, in the mid-1920s, was still essentially rural. Most of the shops were run by traders who had long associations with the city: Mr. Orchard was the bootmaker, Mr. Taylor the seedsman, and a Mr. Bing retailed stone-bottled ginger beer and lavender soap. He brought seventeen acres of fragrance to Grove Ferry, where he cultivated on the banks of the river, carrying the crop into a distillery at the edge of the town, quite near an old-fashioned ford. Cultivation ceased around 1930, according to a local informant, although thirty years later, an old Mitcham grower indulged himself with five acres of lavender in the coastal environment of Cliffsend, Kent, cutting the lavender with saw-edged sickles, to supply coach-loads of tourists.[89]

In a sylvan stretch between Wimbourne Minster and Poole

Harbour in Dorset, Charles Rivers Hill broke up the sandy soil before he carefully cleared and fertilised it. Rivers Hill was a talented violinist, ventriloquist, amateur hypnotist and a born experimenter. He even tried planting throughout the winter, in the hopes of securing a succession of crops to supply regular employment.

For a time, all went well. At his Corfe Mullen Farm, sixty acres of exotic herbs: mint, rosemary, thyme, violets and lavender, waved in the wind. Double white roses hedged the boundaries; and the unused church of St. Andrew served as a distillery, adding, as Rivers Hill aptly remarked, an odour of sanctity to all the other sweet scents.

There was held to be a great future in Dorset lavender: after all, practically valueless tracts of heath were being converted into a productive industry. But the partnership was dissolved at the end of ten years, and produce from the farm gradually diminished. Local people firmly believed that the ghost of a pretty young woman, with two long plaits and a crinoline, remained to haunt the old buildings. She was appropriately known as the lavender lady.[90]

In Bedfordshire, Charles May of Ampthill, brother to one of the founders of Bryant and May, the famous match firm, grew and distilled medicinal herbs in the early nineteenth century. A business partner succeeded him, growing a number of different herbs from his farm in Church Street, Ampthill. Old folk from the district can just remember the time when women picked poppies and gathered lavender.[91]

Lavender was also grown for distillation during the 1950s by monks at Caldey Island, off Tenby, on the coast of Pembrokeshire. There is something satisfying in the knowledge that a long, monkish tradition for herb-growing was being maintained, and photographs show darkly-cowled figures moving over sandy ground between lavender bushes, set against a romantic Abbey. But distillation never progressed beyond the experimental stage, and labour costs were found to be uneconomic. Since 1978 the Brothers have imported flowers in bulk from France. They use them for pot-pourri and sachets to sell, alongside a limited number of fresh lavender bunches, at their three seasonal shops.

But second place to Mitcham as the home of lavender always belonged to Hitchin.

19
Hitchin's Boom

There is no lavender grown in Hitchin today. True, there are ten or twelve acres near St. Ives that help to keep Hitchin stills in action. But the Georgian High Street shop that formed an outlet for lavender products until 1961 has been replaced by a modern Woolworth's, and with it has gone the cobbled yard where local women and children passed summer evenings chopping stalks from blooms. Gone, too, are the thirty-five-odd scented acres eulogised by one Victorian visitor, who vowed he would never forget the impression made by: "Coming suddenly upon a great expanse clothed with lavender in full blossom, the plants set in serrated [sic] ranks, cultivated to the extreme height of floral productiveness, and with no other tint to break the broad, level rich blue-grey blaze."[92]

Why Hitchin should be the place to rival Mitcham for its lavender is a fair question. Perched on the edge of the Chilterns, overlying light gravelly subsoil, the town lies in a ring of hills hard upon the northern borders of Hertford. Until this century the county was largely agricultural. Market garden crops still flourish in the Lea Valley, and agriculture remains the single most important industry. Several books on the history of Hertfordshire give 1568 as the date lavender was first cultivated in the county; but the Encyclopaedia Britannica (at least until 1953), which may be their common source, only hints that it was 'often said' to have been grown in Hitchin then. Even this suggestion, however, has been dropped from the latest editions. If there are references to lavender in early local records, I am afraid that I have been unable to find them.

The first tentative commercial endeavour began back in 1760, when a chemist called Harry Perks established a pharmacy. In 1823 his son, Edward, also a chemist, planted lavender; as it were sowing seeds for the colourful industry that was run by only three different families in nearly 150 years, earning them a series of international awards and turning the town, during the summer months, into a tourist attraction. By 1840, when Edward's son, Samuel, took over the business, they were cultivating thirty-five acres of land on sunny slopes to the north and west of the town, yielding over two thousand gallons of lavender water each year. The acreage, however, was not large by Mitcham standards.

Growing lavender is always a bit of a gamble. Things went smoothly, until a severe frost killing off most of the young plants was followed by an epidemic of shab. Imagine Perks's consternation when, just as his plants were about to flower, they began to wither away. In an attempt to control the disease he transplanted his crop into fresh soil and changed his method of propagation. Instead of pulling off six- to nine-inch slips, only very small green cuttings were taken from healthy plants. It was successful. Two years later, awards and prizes began to roll in. From London, Paris and Philadelphia, accolade followed accolade; well-earned return for all the hard work that had gone into the little industry.

Cropping began when the plants were two years old; although, as in Mitcham, they were not considered to be at their best until the third year. In the fifth, just as they were beginning to run to wood, the whole lot would be uprooted to provide a gigantic November flare. "So big are the bonfires and so powerful the aromatic fragrance given thus to the winds, that Hitchin to its last court and alley is for days together a town of sweet smells."[92]

Around the first week of August, cottage women spent long days cropping the bushes, and binding the stalks into twenty-two pound sheaves, for sixpence a day and free lemonade. The bundles were carted down hill to a small courtyard at the back of the chemist's shop where they were prepared for the still by a merry brigade of women and children squatting in groups, rescuing blossoms from the severed portion after cutting off as much of the stalk as possible, because Perks believed this produced the finest oil.

Ladies sprinkled Perks's Lavender Water on their bodies, poured it into their baths, and hung little embroidered muslin bags of the dried bloom in their wardrobes, whilst gentlemen dabbed lavender water on their faces and shaved with Perks's Lavender Bloom Shaving and Toilet Soaps, one of which, tinted a delicate mauve, was called the 'butterfly brand'. Ladies and gentlemen alike rubbed Lavender Charcoal Dentrifice on their teeth – a powder made from the flowers and wood of lavender before the appearance of toothpaste. "A new preparation", wrote the ingenious director, "that the proprietor believes has never before been introduced to the public".

Samuel Perks, Edward's son, eventually took over the draper's

Perks and Llewellyn's cobbled yard in Brand Street, Hitchin, with Mr. Llewellyn standing, centre.

Miss Lewis's private museum at Hitchin. An illustration (in colour) of the stained glass window can be seen on page 16.

shop next door, and went into partnership with Charles Llewellyn in 1877. As "Perks and Llewellyn", the firm established a countrywide reputation. Children sipped Perks's summer beverages in variety; Hitchin Aromatic Ginger Beer was said to be unequalled for its quality, purity and flavour. Even sheep-dips were prepared at the back of the shop in Hitchin High Street.

All Perks's lavender water bore a coloured label showing a lithograph of a half-timbered cottage called Mount Pleasant. It stood on the crest of a large lavender field, and was often known in the neighbourhood as Lavender Cottage.[93]

Hitchin's lavender reached its peak late in the nineteenth century: difficult times were to follow. Sylvia Pankhurst was not impressed by working conditions in the business. "Lavender water is falling on evil days", she remarked. "People nowadays prefer more modern perfumes with strange sounding names. Hitchin is the largest growing centre but even here a decreasing number of casual workers, both men and women, are employed. The women are mostly of the very poorest and the wages are very low."[94]

The intrinsic vicissitudes of lavender growing; competition from continental Eau de Cologne; and high government taxes on spirits, were beginning to weigh heavily. But Edward Perks's successors kept the firm going until 1961; and fortunately, Miss Lewis, the last proprietor, has set up a private pharmaceutical museum to commemorate the industry. In two rooms, lined with mahogany cases, there is a fascinating display: receptacles, prize-winning trophies, and samples of Perks and Llewellyn's products, recreating the appearance and atmosphere of a Victorian pharmacy.

In the door, there is a beautiful stained-glass window, designed in yellows, reds, blues and rich mauves. At the top, on the left side, is a piece of belladonna; on the right is digitalis, or foxglove; and, in the centre, there is a leech – the last ones ordered from the firm went to a hospital in 1930. Below the leech is a glass vessel for lavender oil; inside a petal-shaped fragment is drawn a lavender plant; and, on the left of it, is a bunch of lavender, cut through with a sickle. There are bees and butterflies, and dried lavender on a sieve. Everything is symbolic of the old lavender days.

There was another firm of growers and distillers in Hitchin. Twenty-two years after Sam Perks first cultivated lavender commercially, a company was founded by William Ransome, son of a long-established Quaker family: millers and farmers who came from Norfolk. William Ransome & Sons cultivated lavender but they did not specialise. To their headquarters in Bancroft (a continuation of Hitchin High Street) trundled carts of belladonna, squirting cucumbers, aconite, lavender and henbane. Sunburnt women arrived with hordes of village children, pushing hand-carts and wheelbarrows; with donkeys; sugar-boxes on wheels; and sacks, and even aprons, full of dandelion roots, buckthorn-berries, poppies, hips and hemlock. It was probably diversification that helped the firm to prosper and march boldly into the twentieth century.

By 1850 William Ransome was already known as a cultivator and distiller of lavender and peppermint oils, and manufacturer of galenicals (herbal preparations) and, like Perks, his work was sufficiently successful to earn him public recognition.

At the Ransome farm, near St. Ives, a variety of herbs was still cultivated as and when required. In 1967, much of their distilling equipment at Hitchin was a century old: tincture jars made by Doulton & Watts (latterly Doulton & Co.); bright copper evaporators with iron legs; digesters, stills and percolators. The director at that time was fourth in line, from father to son, to tread through the archway of their premises in Bancroft.[93]

Mrs. Barker and Mr. Hyde removing leaves and stalks from lavender during the 1960s at Perks and Llewellyn's.

Mrs. Barker, employed by Perks and Llewellyn, seen here with a lavender cart, 1961.

20
Mostly in Norfolk

At one of the world's largest perfume compounding sites in the lap of the North Downs, lavender oil is bought in and stored in 5-ton tanks until it is ready to be used. Then it is drawn off in precise quantities by compounders buzzing around on little electric trolleys. Everything is highly automated. Once materials have been blended they are checked, and drummed in violet containers for transit all over the world.[95]

Lavender oil is not only used in traditional toiletries, but also in air fresheners, washing powders and so on. As western standards of living spread, the demand for perfumes continues to increase. There is, for example, an awakening market for lavender products in developing Africa and South America. But the bulk of the oil used in this country is imported. There are only three parts of Britain where it is grown commercially: Norfolk, St. Ives and Scotland's Banchory.

In North Norfolk, where more than 90 acres of lavender are ruffled by offshore breezes, the largest acreage in the country is centred around the small town of Heacham. The industry stocks gift and chemist's shops with locally grown and packed soaps, talcs, perfumes and bath products; as well as supplying Yardley's with their quota of British oil.

The enterprise began in the early 1930s when Linnaeus Chilvers, a well known nurseryman, joined forces with a local landowner. Initially, the crop was sent to a distillery in Suffolk for oil extraction, but in 1936 Yardley Ltd. located two French stills, which the partnership acquired, and lent them a third. All three are now in the possession of Norfolk Lavender Ltd. A chemist supplied a special formula for the perfume, dating back to King George IV, and they were away.

The natural surroundings: proximity to coast; light soil; high alkalinity; and relatively low rainfall are ideal for lavender. Planting chiefly the prolific Giant Blue; a selected stock of Dwarf Munstead; Old English; and some large-headed Royal Purple, Mr. Chilvers cultivated with an old horse and hoe, eventually expanding his original two acres to about a hundred.

Forking lavender into the still at Fring distillery, N. Norfolk, 1975.

Just as at Hitchin and in Mitcham, the post-war period was hard hit by shab. But when it was realised that the fields were in danger of extinction, a Yardley director visited lavender holdings all over the world, collecting seeds and plant material to replenish the stock. Of the three strains he selected, one is still grown for Yardley's oil, while strains selected by Norfolk Lavender supply their own products. The typical fresh aroma of the latter results from selecting strains whose oil, refined and processed, smells like the flowers in the field.

At present, seven strains are grown for harvesting, but there is a continuing programme of experimental distilling of specially selected seedlings, with a view to extending the distilling season. Early and later flowerers are sought and tested for yield and quality of oil. The ester content of Norfolk lavender strains is high, between forty-six and fifty-two per cent. Year after year the crop flourishes; a striking sight in full bloom, when row on row of misty blue meets the limpid blue of the sky at the horizon, poppies tinge the edge of the fields, baby partridges scuffle out of the undergrowth and bees sound drugged with delight.

Until 1964, women gathered-in the harvest, cutting off the inflorescence, with a short length of stem, before it was sacked and transported to the distillery. This method of harvesting was picturesque, but latterly uneconomic, and has since been replaced by a machine which combs-in the flower spikes towards a cutting blade, then carries them up a conveyor belt into sacks, which are taken by tractor for immediate distillation. The first custom-built cutter was made from an old 'Bean' cultivator frame with a small Ford eight-horsepower engine to run it; and a knife and elevator operating on a very strong spring. Its successor incorporated a hydraulic ramp for raising and lowering the lift, and the third was completely hydraulically operated, with an air-cooled diesel engine.

That lavender thrives on hard cutting has long been recognised. A seventeenth century gardener, Leonard Meager, knew that "Lavender ought to be cut even and handsome so soon as you have your crop off, and if it cannot be made smooth and handsome at one cutting, you may cut it a second time";[96] and American growers regularly cut their plants down to two or three inches from the ground.

In fact, so favourable is the plant's response to mechanical harvesting in regular, beehive-shaped rows, that part of one field

Tapping oil from a condenser at the Norfolk Lavender Company Distillery, Fring, 1975.

belonging to the Norfolk Lavender Company is still in production after more than 20 years. Little cultural attention is needed after initial sub-soiling to break up the hard pan and encourage drainage. For the first two years after planting, tractor cultivation and hand-hoeing are necessary; then regular spraying, with occasional tractor cultivation and spot-spraying of obstinate weeds such as bindweed and ground ivy. From time to time it has been suggested that yields of oil can be increased by fertilizing; but tests found it of only marginal benefit the first year, and of none whatsoever thereafter.

In the new distillery an electric boiler keeps the three copper stills in action. The flowers are tipped helter-skelter on the floor, then forked in, trampled down and distilled. Five or ten minutes later oil begins to be released, and at the end of an hour about two pints of oil have been collected.

Under the present directorship of Mr. Henry Head, the froth of blue fields, bordered with rows of African marigolds, outside Caley Mill, makes a pleasant spot for an August tea. Generous skies lend a peace to the picture, and trout sun themselves in the river Itch, which used to operate the wheels of an oil grain mill.

Harvesting by machine in north Norfolk.

Apart from the Norfolk fields, there are now only about five acres of commercial lavender; and these are at Banchory in Aberdeenshire. A Shetlander set up his pharmacy here just after the last war. The story began in a kitchen, and progressed to a garage, before moving to a small factory. It is not easy to miss: rather a stark little place at the east end of the town, where Dee Lavender is made, together with hand lotion, cleansing cream, vitamin foundation-cream and after-shave.

The business is run as a combined tourist attraction, cottage industry and scientific endeavour, selling largely to visitors, who land on the doorsteps in coach-loads from March until October for a lingering smell of Scotland. The harvest is hand-cut by family and friends, and visitors are able to watch the different processes involved in the preparation of cosmetics, sometimes conducted round by one of the two directors, who are also production chemists and salesmen.[97]

21
A Common or Garden Plant

"To garnish borders" has always been lavender's privilege. Monks used it in this way in the Middle Ages, and we still do so today. Leonard Meager included it among "several herbs fit to set Knots with, or edge borders to keep them in fashion", adding, "Lavender as it may be kept, will be both low and handsome".[96]

William Lawson was another seventeenth century gardener whose advice has stood the test of time, and he coupled roses and lavender because they both yield "profit and comfort to the senses".[19] "Rosewater, Lavender", he wrote: "the one cordial the other reviving the spirits by the sense of smelling." Not only in pot-pourri is this combination superb; but in the garden a soft fuzz of lavender helps to hide the leggy stems of hybrid tea-roses throughout the winter. With all kinds of roses, lavender goes equally well: nothing sets off their warm colours and flowing contours better than spiky silver foliage.

One of lavender's special assets is adaptability: and this accounts for the fact that it survived generations of changing garden fashion. When Louis XIV's grand French style arrived in Britain: when anyone who was anyone wanted long military vistas and ornate parterres, lavender was used in pattern making. Then came landscape parks, sweeping away hundreds of floral chessboards in favour of tree-clad knolls, limpid lakes and undulating lawns.

Along with other herbs, lavender retired to a corner of the cottage garden, to flourish amongst giant poppies and ruddy pickling cabbages, lowly violas spreading gooseberries and a hundred other plants, quietly breathing fragrance until its early nineteenth century reprieve. Yet there were some gardeners who continued to grow flowers throughout the eighteenth century. Miller's Dictionary (1731) is a good guide to plants and horticultural methods of the period, and it gives detailed instructions on how and where to grow lavender. Unlike the majority of garden writers, Philip Miller rebelled against the plant's traditional use in edgings and borders, "for which purpose they are by no means proper, for they will grow too large for such designs; and if they are cut in very dry weather, they are subject to decay". Since the dwarf lavenders we grow today were not available in the 18th century, Mr. Miller definitely had a case.

Such reservations, however, do not seem to have worried the Surrey school of gardeners, who countered the problem of size by selecting small varieties. In fact, lavender was a great favourite with Miss Jekyll, William Robinson and Mrs. C. W. Earle; partly, no doubt, because it is suited to the light, greensand of their gardens. "Best among all good plants for hot, sandy soils are the ever blessed Lavender and Rosemary, delicious old garden bushes that one can hardly dissociate", wrote Miss Jekyll in *Home and Garden* (1900). In another of her books, she suggests planting white and purple clematises so they can be trained freely in and out of lavender.[98] This would be a similar planting to one I have seen at Great Dixter in Sussex, where Christopher Lloyd threads a grey-leaved *Senecio* with two kinds of clematis. The large, indigo-blue *C. praecox durandii* and the spidery, soft-lavender-coloured *C. jouiniana*.

Like Philip Miller, Mr. Lloyd is refreshingly honest about lavender's shortcomings; including an extended off-season, lasting from August flower-fade until a flush of new growth arrives the following May. Nevertheless, he ranks it as one of 'our most cherished eye-sores'.[99] Moreover, the shrub's nine-month dormancy is understandable if you consider the winter conditions in its native climate.

Lavender grows wild on the barren hills of Mediterranean lands; especially in Southern France, where, years ago, it invaded impoverished farmland, abandoned as increasing industrialisation drew the rural population into cities. Here it survives bitter winter storms raging over the snow-covered mountains; and heavy spring rainfalls; coating the slopes in July and August with unbelievably delicate shades of blue.

Obviously, when cultivating lavender in the garden, it pays to simulate its natural habitat as far as possible. Since its stems are strong and wiry, the plant is not troubled by the wind; on the contrary, it actually chooses to grow in open spaces where the air circulates freely. What it dislikes more than anything else, is poor drainage and dull weather. Above all, it demands sun.

Today, you can choose a lavender to suit the needs of your garden.

Perhaps the most popular of all, and ideal for edging in a limited space, is 'Hidcote'. Its rounded cushions and vital purple make a pleasing contrast with red brick or stone paths. 'Hidcote' was one of the plants discovered in Major Lawrence Johnston's beloved Cotswold garden: its history is not known; perhaps Major Johnston brought it over from France. At any rate it seems largely to have superseded the larger, more lavender-coloured, 'Twickel Purple' (bred, presumably, at Twickel Castle in Holland); but 'Munstead', selected by Miss Jekyll at Munstead Wood near Godalming, is still very much in circulation. It is the same size as 'Hidcote': one and a half to two feet and equally free-flowering: though again, a paler colour – in fact, more of a blue. Unlike the majority of garden forms, neither of these early-flowering dwarf sorts shows obvious signs of hybridity.

There are two pinks, 'Rosea' and 'Loddon Pink', which differ only in the leaves being greener, and the flowers more purplish, in the former. 'Loddon Pink', which came from Mr. Tommy Carlisle's Nursery at Twyford in Berkshire, has a certain prettiness, although, admittedly, it is inclined to look wishy-washy beside the blues. I have no hesitation, however, in recommending the whites. Either *alba,* at 3 feet, or the diminutive 'Nana Alba', which makes an endearing tufted sphere, no more than six inches high. It is ideal for walls, window boxes or the conservatory.[100] 'Hidcote Giant', a coarse but very fragrant hybrid is of interest because it bears resemblance to the lavender grown in Mitcham; and 'Grappenhall' is as good as any of the larger forms, about three feet with greyish leaves and strong, loose spikes; looser than those of *L. angustifolia.*

You may notice that Old English lavender is still called *spica* in the majority of nursery catalogues, although Hillier's *Manual of Trees and Shrubs* (1978) gives preference to the new botanical name, *angustifolia.* In most catalogues, too, the broad-leaved *vera* is referred to as 'Dutch lavender' despite the fact that the plant called 'Dutch lavender' throughout history is *L. latifolia,* and botanists say that *vera* is not a distinct cultivar, but simply a synonym for *L. angustifolia.*

We have already mentioned that not all lavender species are completely hardy. The cut-leaved *L. dentata,* requires some protection, and even then it succumbs in a bad winter; so does the slightly smaller *L. pedunculata,* with deep purple corollas peeping between violet-purple bracts; and the short, very fragrant bright violet *L. lanata*. French lavender, too, comes into this category; and though historically prominent, it is no longer seen as often as it merits. It would be the one to choose as a pot plant for Easter.

But even the Old English lavender is becoming less common, now that the dwarf varieties are so much in vogue. I must say that I miss the slightly unruly, slightly headstrong, clumps of grey, smothered with flowers. It is, moreover, dead easy to propagate. Mrs. Earle took cuttings in May, 'puddling' them in, which means first wetting the ground thoroughly, if the weather was dry. This is a bit earlier than is generally recommended. Two-inch heads of firm young growth can be rooted in a propagating frame from June until August; or fully ripe growth in September and October.

A hedge of Old English lavender tumbling over the front wall, was one of the things that attracted us to our present house, although it was, on closer inspection, a mass of contorted stems. The autumn we moved in, I pulled off some forty or fifty cuttings, dipped the ends in hormone rooting powder, stuck them in a plastic tray of John Innes seed compost, wrapped the whole thing in a polythene bag, and left it in the corner of a leaky conservatory. On a couple of occasions I shook condensed water from the bag, but that was all the attention the cuttings received until the following spring; when every one seemed to have rooted – a credit to lavender's powers of propagation. At the end of the year I grubbed out the old hedge, and, six months later, planted one and a half year old, nicely rounded bushes. Two years from scratch, we had a reasonable hedge again.

Lavender is versatile. It is grown in big or small gardens, in herb plots and in municipal parks, in town, country or suburbia. At Cambridge University Botanic Garden there are two long hedges of lavender evenly interspersed with plants of rosemary, leading up to the front of an early Victorian lodge. Offset by plumes of pampas and whitish stone walls it is a very pretty sight. Another pleasing and unusual feature is the forecourt at Wilton House, where pleached limes enclose four flower-beds surrounding a fountain. The simple formal beds, backed by taller shrubs, are filled with lavender in several shades from pink to purple. The late Sir Frederick Stern (who enlightened the horticultural world about gardening on chalk) grew dwarf lavender and white regal lilies together. The shrub protects the lily from late frosts and they both flower at the same time. But my examples could be extended almost indefinitely.

I end the garden chapter with a piece by Charles Estienne, the most delightful of sixteenth-century writers, who clearly loved his garden and his lavender. His *Maison Rustique* is as much revered in France as Gerard's *Herball* is in England.

The most pleasant and delectable thing for recreation belonging unto our farmes is our flower gardens . . . to behold faire and comely proportions, handsome and pleasant arboures and as it were closets, delightfull borders of lavender, rosemarie, boxe, and other such like; to heare the ravishing musicke of an infinite number of prettie small birdes, which continually day and night doe chatter and chant their proper and naturall branch songs upon the hedges and trees of the garden: and to smell so sweet a nosegaie so neere at hand.[101]

22
Sachets and Parasols

In Elizabethan times, lavender oil was used to preserve and remove the musty smell from leather, and the stems were used for burning.[102] There is no reason at all why the latter should not be done today, preferably using the Old English variety. If placed on lengths of material in a dry airy room for several weeks, the flowers can be rubbed off, and the stems tied into bundles to smoulder like sticks of incense. Miss Jekyll preferred old bushes for pictorial effect and young ones, not more than four or five years old, for cutting.[98] If lavender is to be kept and dried, she said, it should be cut when only a few of the blooms are out on a spike. If left too late, the flower shakes off the stalk too readily.[103]

We know, too, that bags or sachets were first used in chests and linen-presses, partly to counteract the rank smell of soap before it was perfumed; and sachets are still made, cut and sewn by my children just as their great-grandmother must surely have done. Lavender sachets have never really left the pages of the more homely women's magazines.

Another absorbing pastime is weaving the stems into bottle shapes. When the plant is in bloom, an uneven number of stalks, nine or eleven, should be snipped off and stripped of their leaves. Thread must then be tied around the bunch just below the flower heads and the stems bent gently back, forming a cage with the flowers inside. A length of mauve ribbon, woven in and out of the stems to the appropriate length, finishes the bottle or 'faggot', a pretty present for anyone's clothes drawer.[104]

In her book on herbs, Rosetta Clarkson describes a number of ways of using lavender: for instance, lavender-filled pads or bags for scenting linens, or padding for coathangers, or sprigs attached to Christmas cards.[105]

Simple fans can be made by laying a number of stalks of equal length on a fan-shaped piece of organdie, covering them with another layer of material, and keeping them in place with rows of lavender-coloured silk thread; and the more elaborate lavender parasols, like those sold by French peasants on the Riviera, might be attempted by the nimble-fingered. They take about 136 heads of lavender, and are a bit more complicated to make. First, bind the heads firmly, about three-quarters of an inch below the tips of the flowers, to a stout ten and a half inch length of wire, then bend the stalks back over the flowers as for a lavender bottle, lacing them in position with a half-inch ribbon to about two inches.[106]

Nine stems, chosen for the frame of the parasol, should be drawn towards the central wire and bound to it with a finer wire to form a handle. This is bent over in a crook at the top, bound in ribbon, and fastened with a small bow at one side. The remaining outer stalks must be cut off evenly all around at three and a half inches from the point where they are turned over, and two and a half yards of soft ribbon, one and a half inches wide, can be used to make a frill at the top. The ribbon should be gathered up on a thread, pushing the gathered edge down between the stems into about ten loops, which are stitched in place. The projecting end of the wire, which forms the point of the parasol, can also be covered with a small piece of ribbon.

Lying on my desk is a pert little wooden-headed lady in a lace bodice and tartan skirt. Only three and a half inches high, she smells sweetly of the lavender sprigs enclosed in her petticoat. My 'Highland Garden' lavender doll is hand made from the Cottage Garden Collection. Provided with a loop to hang over a coat hanger, she will no doubt drive away "fretting moths". A cheerful soul, she will also waft the images of sun and sky into the house, symbolising the romance of Surrey's flower fields which are now only a memory.

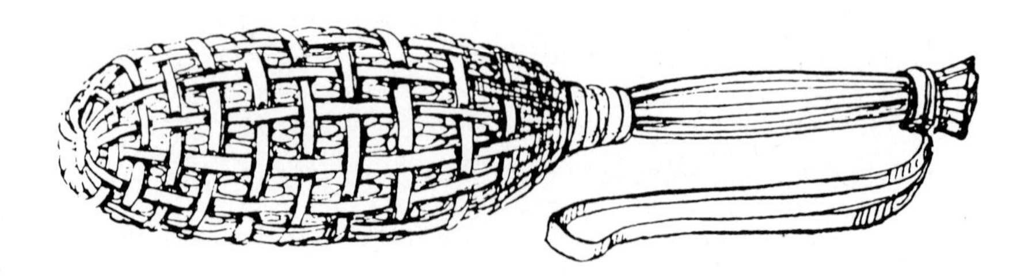

Lavender faggot.

References

1. *Daily Mail*, 1914.
2. William Turner: *A New Herball*, 1551-68.
3. Quoted in: *Herbal Delights* by Mrs. C. F. Leyel. Faber & Faber, 1937.
4. William Langham: *The Garden of Health*, 1579.
5. Izaak Walton: *The Compleat Angler*, 1653.
6. John Keats: *The Eve of St. Agnes* (poem).
7. Article by M. L. Green in *Kew Bulletin*, 1932.
8. John Gerard: *Herball*, 1633.
9. Richard Folkard: *Plant lore, legends and lyrics*. Low, 1884.
10. Dioscorides: *The Greek Herball;* edited by R. T. Gunther. R. T. Gunther, Magdalen College, Oxford, 1934.
11. Pliny: *Natural History;* trs. by Goodyer, 1634.
12. Quoted in: Roy Genders *History of Scent*. Hamish Hamilton, 1972.
13. Hildegarde . . . *Patrologia Latina*, ed. Migne: vol. 197 from *Liber subtilitatum diversarum naturarum creaturarum*, 1150-60.
14. Anon: . . . *an Herball*. Published by Rycharde Banckes, 1515.
15. Mrs. M. Grieve: *A Modern Herbal*, 1st ed., Cape, 1931.
16. Herbert Schöffler: *Beiträge zur Mittelenglischen Medizin-literatur III*, Anglistrische abtilung [English edition], heft 1, Halle a.s. 1919.
17. Warren R. Dawson: *A Leech Book, or, Collection of Medicine Recipes of the Fifteenth Century*. London, 1934.
18. Edmund Spenser: *Muiopotmos;* or, *The Fate of the Butterfly*. (poem).
19. William Lawson: *The Country Housewife's Garden*, 1617.
20. Thomas Tusser: *Five Hundred Points of Good Husbandry*, 1573.
21. Audrey Wynne Hatfield: *Pleasures of Herbs*. Museum Press, 1964.
22. Gervaise Markham: *Way to get Wealth*, 1631.
23. W. J. Stearn: Article on Gerard in *Dictionary of Scientific Biography*, New York, 1971.
24. John Parkinson: *Paradisi in Sole, Paradisus Terrestris*, or, *A Garden of Pleasant Flowers*, 1629.
25. Nicholas Culpeper: *Complete Herbal* (1815 edition).
26. Mrs. C. F. Leyel: *Herbal Delights*, Faber & Faber, 1937.
27. Dorothy Hall: *A Book of Herbs*. Angus & Robertson, 1972.
28. Hans Flück: *Medicinal Plants*. Foulsham, 1976.
29. *Harpers and Queen* magazine, September 1977.
30. Jacqueline Heriteau: *Pot Pourris and other Fragrant Delights*. Lutterworth, 1975.
31. Mrs. C. W. Earle: *Pot-pourri from a Surrey Garden*. London: Smith, Elder, 1900.
32. Constance Isherwood: *A Bunch of Sweet Lavender*, Hitchin, Paternoster & Hales, n.d. Illus. brochure on lavender industry. (Copy in Miss Lewis's private Lavender Museum in 1977).
33. Sir Matthew Hale: *Primitive Origination of Mankind . . .*, 1677.
34. Sir Hugh Platt: *Delight for Ladies*, 1609 (repr. 1955).
35. Donald McDonald: *Sweet Scented Flowers and Fragrant Leaves*. London, Low, 1894.
36. Robert Tyas: *Language of Flowers*. London, Routledge, 1869.
37. Baron Frederic de Gingins-La Sarraz: *Natural History of the Lavenders*, tr. from French by members of the Herb Society of America. Boston, 1967.

38. William Shenstone: *The Schoolmistress,* 1742 (poem).
39. Poem by M. E. Davies in *Hitchin Grammar School Magazine* c.1960 (copy in Miss Violet Lewis's private Lavender Museum, Hitchin, in 1977).
40. *Diddle diddle,* or, *The Kind Country Lovers,* quoted in: Iona & Peter Opie, eds: *Oxford Dictionary of Nursery Rhymes,* O.U.P., 1951.
41. *Songs for the Nursery,* 1805; quoted in Opie *op.cit.*
42. S. J. Adair Fitz-Gerald: *Sweet Lavender;* article in: *Lady's Pictorial,* August 8th, 1914.
43. Gammer Gurton's Garland, 1784; quoted in Opie *op.cit.* (40 above).
44. Quoted in Opie, *op.cit.* (40 above).
45. John Betjeman: *South London Sketch,* 1944 (poem).
46. G. W. Septimus Piesse: *Art of Perfumery.* London, 1855.
47. Alfred Heales: *The Records of Merton Priory . . . Surrey,* 1898.
48. John Murray (publisher): *A Handbook for Travellers in Surrey,* various eds, c.19.
49. James Malcolm: *A Compendium of Modern Husbandry . . .,* 3 vols. London, Baldwin, 1805.
50. James Thorne: *Handbook to the Environs of London,* Pt.2, 1876.
51. Surrey Record Society: *Mitcham Settlement Examinations 1784-1814;* ed. Blanche Buryham. S.R.S., 1973.
52. John Hassell: *Picturesque Rides and Walks,* vol. I. London, 1817.
53. Rev. Daniel Lysons: *The Environs of London,* vol. I, 1792.
54. Edward Walford: *Greater London,* Vol. II. London, Cassell, [1883-4].
55. Quoted in Walford, *op.cit.* (54 above).
56. *Pharmaceutical Journal and Transactions,* Vol. X, 1850-51, p.116.
57. James D. Drewett: *Memoirs of Mitcham, in* Parish Magazine, 1923.
58. Pigot's Directory, 1825.
59. Robson's Commercial Directory for 1839.
60. Directories in Mitcham Library's Local Collection (various titles), as quoted by Jane Bulmer (see 74 below).
61. *Pharmaceutical Journal and Transactions,* Vol. X, 1850-51.
62. Photographs in Sutton Libraries' Local Collection.
63. Mr. K. A. Pryer: Letter in Sutton Libraries' Local Collection: Lavender file.
64. The Distillation of English Essential Oils; Messrs. J. and G. Miller, Mitcham, Surrey: reprint from *The Gentleman's Journal and Gentlewoman's Court Review,* May 16, 1908.
65. *Girl's Own Paper,* May 23, 1891.
66. Inventory of Valuation, Little Woodcote Farm quoted in *Surrey's Fragrant Harvest* by S. H. Chalks in *Surrey Life,* September, 1973.
67. *The Cultivation and Distillation of Peppermint;* a three-page typescript by Mrs. Hilda Machell, in Sutton Libraries' Local Collection.
68. *Chemist and Druggist,* 1891. Letter and articles by J. Ch. Sawer, March 7, 14, 21.
69. W. B. Brierley: *A Phoma Disease of Lavender;* in *Kew Bulletin,* 1916.
70. *Chemist and Druggist,* September 12, 1891.
71. Jane T. Bulmer: *A Study of the Growth and Development of the Borough of Mitcham with Special Reference to the Lavender and Herb Trade,* April, 1970. Photocopy of manuscript in Mitcham Library.
72. *Culinary and Medicinal Herbs,* Ministry of Agriculture Bulletin No. 76, H.M.S.O., 1960.
73. Arthur Mee: *Surrey* (King's England Series) [Pre-Second World War edition].

74. *Chemist and Druggist*, September 15, 1874.
75. Tom Francis: Notebook. February 1943. (Typescript in Mitcham Library).
76. *The Daily Chronicle*, August 13, 1910. (Cutting in Mitcham Library).
77. MS in Mitcham Library by F. H. Priest of Messrs. W. J. Bush Ltd., 1953.
78. Charles Hindley: *A History of the Cries of London Ancient and Modern*. London, 1884.
79. *Journal of the Folk Song Society*, June 1919.
80. William Stewart: *Characters of Bygone London*, Harrap, 1960.
81. Typed sheet in Miss V. Lewis's private Lavender Museum, Hitchin.
82. *Mitcham News*, September 10, 1954.
83. Dorothy Davis: *History of Shopping*, Routledge, 1966.
84. *Daily Chronicle*, August 8, 1914.
85. *Mitcham Post*, August 18, 1930.
86. *Chemist and Druggist*, January 16, 1932.
87. *Mitcham News*, July 26, 1958.
88. *Journal of the Folk Song Society*, June 1919.
89. Sir Charles Igglesden: *A Saunter through Kent with Pen and Pencil* Vol. XVIII, Ashford, Kentish Express. 1926.
90. Olive Kent: *The Lavender Farm* in *The Dorset Year Book*, 1956-7.
91. H. G. Tibbutt: *The Herb Farms of Ampthill* in *The Bedfordshire Magazine*, Winter 1947/8.
92. Cassell's *New Penny Magazine*, 1898.
93. W. Branch Johnson: *Industrial Monuments in Hertfordshire*. David and Charles, 1967.
94. *Evening News*, August 26, 1907.
95. Publicity from Proprietary Perfumes Ltd., Ashford, Kent.
96. Leonard Meager: *The New Art of Gardening*, 1697.
97. John Rundle: *Bouquet from Banchory* in *The Scots Magazine*, May 1978.
98. Gertrude Jekyll: *Colour in the Flower Garden*, 1908.
99. Christopher Lloyd: *The Well-Tempered Garden*. Collins, 1970.
100. Kay Sanecki: *The Complete Book of Herbs*. Macdonald, 1974.
101. Charles Estienne: *Maison Rustique*. Paris, 1572; Quoted in: W. Thomas Beach, *Gardens*, Burke, 1952.
102. Elizabeth Jenkins: *Elizabeth the Great*. Gollancz, 1958.
103. Gertrude Jekyll: *Wood and Garden*. London, Longmans, 1899.
104. Beverley Plummer: *Fragrance*, Hale, 1975.
105. Rosetta Clarkson: *Herbs, their cultivation & use*. Macmillan, 1967.
106. *Home Chat*, April 15, 1931.

Index